A series of student texts in

CONTEMPORARY BIOLOGY

General Editors:
Professor E. J. W. Barrington, F.R.S.
Professor Arthur J. Willis

Life of
Marsupials

Hugh Tyndale-Biscoe
Reader in Zoology
Australian National University,
Canberra

American Elsevier Publishing Company, Inc.
New York

American Elsevier Publishing Company, Inc.
52 Vanderbilt Avenue, New York, N.Y. 10017

First published in Great Britain by
Edward Arnold (Publishers) Ltd.

ISBN: 0-444-19558-0
Library of Congress Catalog Card Number: 73-2310

599.2
T 97e

Printed in Great Britain by
William Clowes & Sons, Limited
London, Beccles and Colchester

Preface

There was a time when marsupials were regarded as second class mammals and the main justification for studying them was for the light they might shed on the evolutionary processes leading to higher mammals. This view was grounded in comparative anatomy and enshrined in T. H. Huxley's classification of mammals into Proto-, Meta- and Eutheria.

The past 15 years in Australia has seen a shift towards the view that marsupials are worth studying for their own intrinsic interest and, in some cases, for urgent economic reasons. The outcome of these studies has been the realization that this single order contains a diversity of species far wider than in most orders of Eutheria, and that among these species are anatomical, physiological and biochemical adaptations for particular modes of life quite the equal of their eutherian analogues.

There is in fact only one aspect of marsupial life that clearly distinguishes all marsupials from all Eutheria and that, as the name implies, is the mode of reproduction. This difference in reproduction stems from a profound dichotomy in the early developmental programme of the urogenital system. This must have occurred very early in the history of mammals although, since it leaves no mark in the skeleton, it is not possible to fix the time. Reproduction is really the only area in which one can talk of a distinctive marsupial condition; in most other areas one is considering a species of mammal, which happens to be a marsupial, and there is little value in attempting to make generalizations about marsupials as such.

In this book, therefore, the emphasis is on living marsupials and how they are adapted to their environment, including their adaptation to changes brought about by man. Perforce most of the examples are Australasian species, although the increasing interest in American marsupials has not been ignored. In writing this I have assumed that the reader has some background knowledge of physiology, such as he can obtain from the companion volume in this series, *Principles of Animal Physiology*, by Dennis Wood.

The burgeoning interest in marsupial biology is resulting in a great deal of new and interesting research, much of it by post-graduate students in Australian and American Universities. I have drawn on this work quite a deal; not to have done so would have been to deny much that is of great interest. For allowing me to use their unpublished results, I acknowledge Dr. Patricia Berger, Dr. R. H. Brocke, John Hearn, Dr. Rory Hope, Dr. John Kirsch, John McIlroy, Dr. Marilyn Renfree, Dr. Cedric Shorey, John Winter. I also wish to thank the several people who have read parts of the book and provided many useful comments: Josephine Bancroft, John Calaby, Dr. Robert K. Enders, Dr. Mervyn Griffiths, Sue Johnston, Michael Plaine and Dr. Marilyn Renfree. To Frank Knight I owe an especial debt for his drawings, which with a deft skill, express the ideas I wished to convey. And finally I acknowledge Marina, whose constant help brought this enterprise of ours to a conclusion.

Canberra,   H.T.–B.
March, 1971

Table of Contents

I

Relationships and Origins

RELATIONSHIPS OF LIVING MARSUPIALS TO OTHER MAMMALS

Marsupials are interesting because they represent a separate pattern of mammalian organization of great antiquity and because their distribution in the world today is so disjunct. They occur only in Australasia and the Americas: none occurs in Africa, Europe, Asia or Antarctica and no fossil marsupial is known from any of these continents, except Europe.

To say this is to pose two questions about marsupials. First, is there a true and close relationship between all marsupials that distinguishes them as a taxonomic group from all other mammals? Secondly, if there is, when and how did the present disjunct distribution arise? Both questions have intrigued men's minds for more than 200 years and though the first has been answered in the affirmative the second remains enigmatic.

Development of ideas about marsupial relationships

Among the curios that Pinzon presented to the King of Spain on returning from his voyage to Brazil in 1500, was a female opossum, the first marsupial to be seen in Europe. A few years later Gessner included this new beast in his Historia Animalium as the simivulpa or monkey-fox.

Notwithstanding its resemblance to familiar animals of Europe, the remarkable provision of Nature in the pouch was noted by the early savants, although confused by Marggrar (1640) as the exteriorized womb. Fifty years later Tyson, writing in the Philosophical Transactions of the Royal Society, provided the first accurate description of the Virginia opossum and recognized the unique double uteri and vaginae in the female. Linneaus, however, took little cognizance of this in constructing his Systema Naturae for he followed Marggrar in coining the name *Didelphis* to indicate the possession of two uteri, one internal and the other the external uterus or pouch. Furthermore, he placed the species along with the pig, armadillo, hedgehog and shrew in the Bestiae because of their common possession of sharp teeth.

Little more than a century after Pinzon collected his opossum, the Dutch captain Pelsaert wrecked on the inhospitable Abrolhos islands off the west coast of Australia described in commendable detail another animal with the same remarkable pouch and tiny offspring contained therein, but it seems likely that his report did not reach Europe then, for it is not acknowledged by Tyson. Other navigators were also encountering similarly bizarre animals when they made landfalls on the New Holland coasts—the Spaniard, Torres, the Dutchmen, Volkersen and de Vlamingh, and the Englishman, Dampier.

From the Dutch trading posts specimens of these animals and drawings of them began to reach Europe. Brisson (1762), recognizing their kinship to the opossum, named the New Guinea cuscus *Didelphis orientalis* but Buffon three years later observed the fused phalanges in the hind feet and distinguished them by the name *Phalanger*, from which the family name has since been derived.

The small arboreal animals from Australasia were comfortably accommodated with the American opossums until the return in 1773 of Cook's first expedition. The naturalists on this expedition had been diligent in collecting animals along the east coast of Australia and presented the first specimens of wombats, native cats and kangaroos. John Hunter and later his nephew Home had an opportunity to dissect a kangaroo and recognized the similarities in the reproductive tract to the opossum. Nevertheless, the resemblances of the kangaroos and wombats to rodents in their teeth and habits confused the taxonomists and Cuvier (1795) was so at a loss to decide the issue that he erected an order Pedimanes to accommodate the opossums and native cats (*Dasyurus*) and placed it between the Carnivora and the Rodentia; the kangaroo being placed in the Rodentia.

In 1792 Shaw described the echidna and considered it to be closely related to the anteater *Myrmecophaga*, and Cuvier included it and the subsequently discovered platypus in the Order Edentata, on the grounds that they both lacked teeth. However, with the ever increasing number of animals being brought home to Europe from around the world the older method of classifying animals by their life forms was proving erroneous, and de Blainville (1816) contributed the important new concept of looking for more fundamental similarities in determining relationships. He recognized that the mode of reproduction set marsupials apart from all other mammals, despite their many remarkable resemblances to Carnivora and Rodentia. The character by which he chose to distinguish them was the condition of the reproductive tract, while retaining Linnaeus' name; in marsupials it is double, hence Didelphes and in the others it is single, hence Monodelphes. Within the marsupials he recognized two groups, the carnivorous and the rodent-like forms. The echidna and platypus he grouped with the marsupials although by this time their distinctive features were recognized by the name Monotremata indicating their possession of a single cloaca for the urogenital passage and rectum, as in birds and reptiles. Although the name remains, this is an inadequate criterion of distinction from marsupials, as the tyro marsupial biologist soon discovers when he attempts to take a vaginal smear and gets a rectal one instead.

When de Blainville revised his classification in 1834, he recognized the distinctness of the monotremes, but chose the same criterion as he had used before. The oviducts in the echidna and platypus resemble those of the bird or reptile in being quite separate, and so he termed the monotremes the Ornithodelphia, whereas for the marsupials, in which the oviducts are partly united in the vaginal region, he retained the term Didelphia and all other mammals he included in the Monodelphia.

His criteria have proved to be good ones, for subsequent work has confirmed the distinctness of these three groups of living mammals. What confused people before him were the quite remarkable convergences to be seen between marsupials and monotremes on the one hand and eutherians of similar life pattern on the other. These have often been remarked[142] and need only be summarily mentioned here: the wombat and the woodchuck, the kangaroo and antelope, the marsupial mole, *Notoryctes typhlops*, and the eutherian mole, the thylacine and the dog, the Tasmanian devil and the wolverine, the water opossum and the otter.

No evolutionary overtones are implicit in de Blainville's classification but in the years that followed it was discovered that the Ornithodelphia laid eggs and it was known, of course, that marsupials bore very small young. Other affinities of the Ornithodelphia with reptiles led T. H. Huxley to suggest that they and the marsupials represent earlier, serial, stages in the evolution of the true viviparous mammals, and he coined another set of names to reflect this idea: Proto-, Meta-, and Eu-theria. These several nomenclatures can be compared:

de Blainville	Bonaparte	Huxley	Illiger, Owen
Ornithodelphia	**Monotremata**	Prototheria	
Didelphia	Ditremata	Metatheria	**Marsupialia**
Monodelphia		**Eutheria**	Placentalia

Although de Blainville's grouping has remained, his names have not and the ones in common use are in bold type. Huxley's idea has had a long and baneful influence on the understanding of marsupials and monotremes; it encouraged people to think that by studying these mammals they could ride a sort of Wellsian Time Machine back to the origin of mammals. Even more, that this was the only proper purpose in studying these animals, notwithstanding the clear evidence of convergence to show that they are marvellously adapted to their present environment. Huxley's grandson, Julian, began to redress the balance by emphasizing in his *Evolution: the Modern Synthesis*, that all living animals must be viewed in the context of their adaptations to the present environment, but the old idea dies hard. It had a great impetus, soon after it was proposed, from the studies of Sutherland[266] and Martin[173] on the control of body temperature in several marsupials and the monotremes. These studies showed that the monotremes and marsupials had lower body temperatures than the laboratory species of Eutheria used and were apparently unable to maintain body temperature against an ambient temperature gradient. These results were for many years quoted as showing that monotremes and marsupials represent stages in the evolution of homeothermy or euthermy. Recent studies on the echidna, reviewed by Griffith,[102] and the many recent studies on marsupials to be considered in later chapters of this book, show that this is not so and that both monotremes and marsupials show the full array of adaptations for euthermy and any species can be matched against an eutherian of similar life pattern: where they differ from Eutheria and from each other is in the level at which the metabolic rate is set.

This is not really surprising, for euthermy involves so many interacting adaptations and affects every function of the body so profoundly, that its evolution must have involved a quantum step in Simpson's[254] meaning of the term, and intermediate stages would be most improbable. What is not clear from the fossil record is when euthermy was achieved in the evolution of mammals; whether it evolved among the Therapsida (Fig. 1.1) or was evolved independently in each line that crossed the mammalian threshold. This is arbitrarily defined by the possession of three ear ossicles, and a single dentary in the lower jaw.[222] By this criterion five groups of fossil animals evolved to mammalian status independently; none of them is clearly ancestral to the monotremes, which are unknown as fossils before the late Tertiary. The marsupials and eutherians, however, have probably evolved from pantotheres by a dichotomy now thought to be in the early Cretaceous. Thus the fossil record, the only certain guide to phylogeny, shows that the three living groups of mammals have been genetically separate for more than 100 million years.

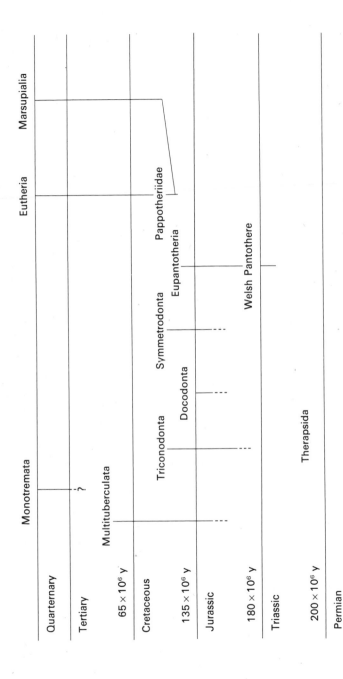

Fig. 1.1 Chart to show geochronologic units and the ranges of mammalian taxa and their reptilian predecessors. The past relationships of the Monotremata are unknown, whereas the Eutheria and Marsupialia have a common ancestry in the Pantotheria. (After Clemens[65])

During that span of time the Marsupialia and Eutheria have each evolved a wide variety of adaptations that the common ancestor could not have possessed and the astounding thing is how similar these are. Not only the anatomical convergences mentioned above but the independent acquisition of ruminant digestion, embryonic diapause, hibernation and counter current exchange in the kidney, to name the more conspicuous ones that we shall be considering later. Apart from the main dichotomy other parallelisms evidently took place within each group as well. If these adaptations were not part of the common heritage their separate evolution suggests that, within the framework of mammalian euthermy, species in a given environment have relatively few options and the same ones arise each time. Awareness of this may dispel the notion that study of living marsupials can shed light on how Cretaceous mammals functioned; on the contrary, the critical study of individual marsupial species within the context of their own life system is likely to be far more fruitful in extending our understanding of how mammals function now.

In the rest of this chapter we will examine the two questions posed at the start. First, the common attributes of marsupials vis-a-vis other mammals and how these vary within the group itself will be considered and then how the fossil history of marsupials helps to answer the second.

Are marsupials a distinct group?

Anatomy

The character that gives marsupials their name is not their most distinctive; very few male marsupials have a pouch and the females of some species only develop one while suckling. So does the female echidna, *Tachyglossus aculeatus*, which is not a marsupial but a monotreme. Nevertheless, it was this organ that arrested Pelseart's attention and subsequent observers were equally impressed by this character, as the prefixes pera-, phascolo-, and thylaco- in many generic names bear witness. However, it is in the relationship of the urinary and genital ducts to each other that marsupials are exclusively differentiated from other mammals. To appreciate this it is necessary to digress, and consider the early development of the urogenital tracts.

The pronephric ducts which grow posteriorly and make contact with the proctodeum become the main urinary ducts in lower vertebrates as well as the conveyers of sperm from the testes. The oviduct grows in association with the pronephric duct and both open dorsally into the urogenital sinus (Fig. 1.2). In the birds and mammals a metanephric kidney develops in the late embryo posterior to the mesonephros as an

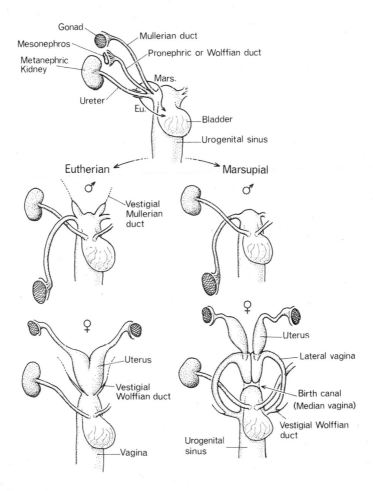

Fig. 1.2 The urogenital tract in male and female eutherians and marsupials, to show how each is derived from a common pattern of kidney and genital ducts at the indifferent stage in the embryo.

outgrowth from the pronephric duct, and subsequently the pronephric duct remains solely as a genital duct. In marsupials and eutherians the urinary ducts have migrated to the ventral part of the urogenital sinus to open directly into the bladder, while the genital ducts have remained dorsal. It is evident that the urinary ducts in their migration must either pass between the genital ducts or outside of them. In all marsupials the urinary ducts pass mesially and in all eutherians they pass laterally. The consequences (Fig. 1.2) of this are that in male eutherians each prone-phric duct must loop around the inside of the ureter to reach the testis, while in male marsupials this loop is absent; and in female eutherians the two oviducts can fuse together in the midline to form a single vagina and uterus, whereas in female marsupials this fusion in the vaginal region is impeded by the ureters and does not occur in the uterine segment of the tract either. Hence among the living mammals the monotremes retain the reptilian form of dorsal outlets for the ureters, the marsupials have two uteri and lateral vaginae and the Eutheria have a single vagina and uterus. In marsupials the two lateral vaginae become united anterior to the ureters (Fig. 1.2) and a median vaginal cul de sac is formed. At parturition a birth canal forms in the connective tissue between this and the urogenital sinus through which the foetus passes. In most marsupials this is transient and reforms at each birth, but in most kangaroos and wallabies, and in the honey possum, *Tarsipes spencerae*, it becomes lined by epithelium and remains patent after the first birth, so that a condition very similar to the eutherian is obtained.

The eggs of Marsupialia and Monotremata share several characters that differentiate them from Eutheria: they are larger, in Monotremata very much larger, and are surrounded by three membranes. The outer shell membrane is retained for most of gestation in marsupials and after the intra-uterine stage in monotremes. In both groups intra-uterine development is minimal and there is a long post-natal phase of develop-ment. From Griffith's[102] study it is evident that the similarities in reproductive processes in the two groups are much closer than was previously supposed and that the laying of an egg, or the birth of an immature young, are not profoundly different, as will be shown in Chapter 2.

In the skeleton marsupials possess a pair of epipubic bones articulating with the pubis, which are not found in any Eutheria but are possessed by monotremes and by at least one extinct multituberculate; the mar-supial humerus has a foramen through the medial epicondyle. Adult marsupials, like Eutheria, lack a coracoid articulation with the sternum but this is well developed in monotremes and occurs transiently in new born marsupials (p. 80).

In the marsupial skull the tympanum is covered by an extension of the alisphenoid bone whereas in Eutheria it is covered by the tympanic bone itself; the dentary has a medial inflection which carries the ptery-goid muscles and the lacrymal bone extends beyond the orbit to cover part of the snout. There are usually four molars and up to three pre-molars in each jaw, in contrast to three molars and four premolars in Eutheria. In primitive marsupials there are five incisors in each pre-maxilla and three in each dentary, but in some families the number is reduced. From this it is evident that the Marsupialia share some attributes exclusively with Monotremata and others with Eutheria and few characters are unique to the group itself. But these, especially the reproductive tract, are so particular that they set marsupials apart unequivocally.

Physiology

As mentioned above, Sutherland[266] found 9 marsupials to have a lower body temperature than Eutheria and shortly after Martin[173] reported that the basal metabolic rates of several species were less than one third of eutherian values. Both studies were taken to indicate less effective thermoregulation in marsupials, which were thus considered to lie between monotremes and Eutheria. In recent years several studies have confirmed the low body temperatures (Fig. 1.3) in a wide range of marsupials. As we shall see in subsequent chapters some species have a more or less marked diurnal cycle, but at any time the body temperature is maintained within close limits over a wide range of ambient temperatures.

Martin's observations on the metabolic rate have only recently come under closer scrutiny, the most comprehensive examination being that of Dawson and Hulbert.[75] Their results on 8 species of Australian marsupial are given in Fig. 1.3, together with the results of two other studies, recalculated to conform to their presentation.

The Standard Metabolic Rate is a somewhat theoretical concept of the minimum energy requirements of an animal at rest. It can be approached indirectly by measuring the oxygen consumption under these conditions for a given period of time. If it is assumed that the energy substrate is of mixed composition and has an average respiratory quotient of 0·8, then $1 \text{ cm}^3 \text{ O}_2 = 20 \cdot 1$ J. Now a large animal will consume more O_2 in a given time than a smaller one (Fig. 1.3) but the smaller one consumes more in proportion to its body weight, that is $\text{cm}^3 \text{ O}_2/\text{kg}$. This is because most metabolic activities occur at surfaces and this increases as the square power, whereas weight increases as the cube power of size. For all animals it has been found that the function that best fits this discrepancy between energy metabolism and size is the 3/4 power of body weight or $\text{kg}^{0\cdot75}$. Thus to compare the energy metabolism of different species this conversion is employed.

To maintain a stable body temperature (T_B), euthermic or homeothermic animals expend least energy when the ambient temperature (T_A) is nearly the same as T_B; when T_A is lower, more energy is required as heat, and when T_A is higher, energy is expended in cooling devices such as panting. The thermoneutral zone (TNZ) is the ambient temperature where minimum oxygen is consumed and this is the value

	Species	N	Weight, kg	T_{Air}	T_{Body}	O₂ Consumption cm³/h	S.M.R. kJ/kg$^{0.75}$/day
Dasyuridae	Sminthopis crassicaudata	6	0·014	32·0	33·8	18·7	220·6
	Antechinus stuartii	6	0·037	30·5	34·4	36·6	209·7
	Dasyurus geoffroii	14	1·300	30·0	36·0	500	198·8
Peramelidae	Perameles nasuta	6	0·686	28·0	36·1	322	204·7
	Isoodon macrourus	6	0·880	28·0	34·7	393	208·9
Diprotodonta	Cercartetus nanus	5	0·070	31·0	34·9	60	212·6
	Petaurus breviceps	6	0·128	30·0	36·4	88·7	199·7
	Trichosurus vulpecula	6	1·982	27·0	36·2	625	180·0
	Macropus eugenii	6	4·796	25·0	36·4	1382	205·1
	Megaleia rufa	6	32·490	26·0	35·9	5791	197·1

Fig. 1.3   Resting body temperatures and standard metabolic rate of 10 marsupials, arranged according to family and body size. (Data for *Dasyurus* recalculated from Arnold & Shield,[15] and for *Cercartetus* recalculated from Bartholomew & Hudson[29] to conform to the rest from Dawson & Hulbert[75])

| | Reptiles | Mammals | | | Birds | |
	Lizards	Monotremes	Marsupials	Eutherians	Non-passerine	Passerine
Body temperature	30	30	35·5	38	39·5	40·5
Standard metabolism $kJ/kg^{0.75}/day$	31·4	142·3	203·7	288·8	347·4	598·5
Standard metabolism at 38°C	81·6	259·5	259·5	288·8	301·4	477·1

Fig. 1.4   A comparison of the standard energy metabolism of terrestrial vertebrates. (After Dawson & Hulbert[75])

used in determining the standard metabolic rate of a homeothermic species; it is expressed as $kJ/kg^{0.75}/day$. From the smallest species weighing 13 g to the red kangaroos weighing 32 kg, the values for 10 marsupials in Fig. 1.3 are very similar and average 203·7 kJ. This is about 70% of the standard energy metabolism of a wide range of eutherian species ranging from the mouse to the elephant,[149] in which the average value is 288·8 kJ, and thus confirms Martin's view although not the degree of difference that he measured. In Fig. 1.4 these values are compared to those of other homeotherms and one group of ectothermic reptiles. Two things are evident in this; there is a clear difference between the reptiles and the others, while among the homeotherms each group has a characteristic level and these are positively associated with body temperature. The monotreme, *Tachyglossus aculeatus*, has the lowest T_B and lowest standard metabolism while the passerine birds havě the highest. When all homeotherms are compared at 38°C the differences are less striking, but the higher values for monotremes and marsupials reflect the fact that this temperature is much above their respective thermoneutral zones. However, the real differences between the three groups of mammals do not indicate a phylogenetic series but rather that each group acquired a different metabolic level in achieving homeothermy; and in birds the still higher level is probably associated with the particular requirements of flight.[295]

The lower setting in marsupials may also confer advantages but of a different kind; the lower energy metabolism means lower food requirements and storage reserves in the body last longer during adverse conditions. Conversely, growth may be slower (see p. 99) and neural responses less rapid than in Eutheria or birds.

The lower metabolic rate of marsupials is reflected in several other parameters. The heart rate of marsupials is lower than in eutherians of the same size;[145] the nitrogen requirement of some species is less than in comparable sized Eutheria,[54] the excretion of creatinine, an end product of protein catabolism, is less[95] and thyroid activity is less.[31] These and other aspects related to metabolism will be dealt with in the later chapters, but are mentioned here to indicate that the lower metabolic rate appears to be a criterion that distinguishes Marsupialia from Eutheria as clearly as anatomical criteria and that pervades their physiological functions as homeotherms.

RELATIONSHIPS OF MARSUPIALS TO EACH OTHER

Accepting, then, that marsupials are a single taxon, what are the relationships within this group and how have they been evolved? The latter question is important because different people at different times

have placed emphasis on one or another suite of ch
real relationship it should be reflected by whateve
differentiate the species. The fact that this does not
cates that, in some characters at least, parallel
occurred in more than one group independently. I
is no strict time dependency. Much of the earlier t
on the assumption that time dependency has been in
and that if two species uniquely share a certain character they must be
derived from a common ancestor. The development of new methods for
analysing a wider spectrum of characters has forced the realization that
time dependency has not been invariable, and an understanding of
relationships is only achieved by examining all available criteria together.

The first attempts to classify marsupials were based on the number
and kind of teeth and on the number of digits on the feet. These characters are undoubtedly affected intimately by the mode of life of the
possessors but on the other hand have the advantage of being available
in fossil remains as well. Nevertheless it should be emphasized that the
hard parts of fossils are no more conservative than other characters such
as serum proteins and are as liable to convergent and parallel alterations,
so that apparently close relationships may be illusory.

Anatomy

Dentition

Marsupials can be divided into two sub-orders on the basis of tooth
number. All the living South American species and the Australian
carnivorous species have long snouts bearing a large battery of simple,
sharp pointed teeth; in each ramus there are four molars, three premolars, one prominent canine and four or five incisors in the upper jaw
and three in the lower jaw (Fig. 1.5). These are termed the Polyprotodonta.

The larger carnivores in Australia as well as extinct species in South
America have a reduced number of molars and premolars and this
feature was at one time taken to indicate close links between Australia
and South America in the Miocene. There is, however, no independent
evidence for this link and the more parsimonious view is to suppose that
both groups evolved in parallel in response to the same kind of selective
pressures, as indeed did the quite unrelated eutherian Felidae.

The herbivorous Australasian species have fewer premolars, the
canine may be absent and there are three incisors in the upper jaw and
one large procumbent incisor in each dentary. The latter character is the
distinguishing one for the sub-order Diprotodonta. Again, a superficial

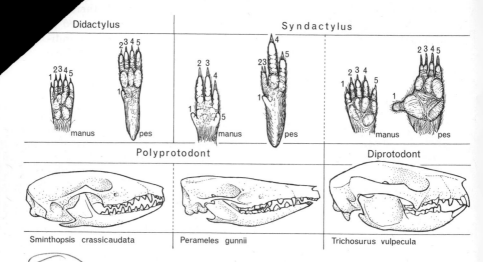

Fig. 1.5 Plantar view of left manus and pes of didactylus and syndactylus species and the skulls of the same species to show the polyprotodont and diprotodont dentition. (Drawings of feet from Wood Jones[4] and skulls from dry specimens)

link is found with one group of South American marsupials, the Caenolestidae, in which the first incisors in the dentary are large and procumbent, but in these species the other posterior incisors are retained as well. The embryological evidence suggests that in the Australian Diprotodonta it is the second incisor which is procumbent, so once again we see a parallel evolution, probably here in relation to change in diet, and again convergent forms are found in the Lagomorpha and Rodentia among Eutheria.

Within the Diprotodonta further distinctions can be made on the basis of teeth. In the phalangerids the molars are squared off by the presence of a hypocone in the upper molars and two additional cusps in the lower (Fig. 1.5) and the cusps are more rounded than in the Polyprotodonta. There is a trend to reduction of the premolars to one and the canine is small or absent. As a result a diastema develops between the anterior cutting teeth and the posterior grinding battery. This pattern reaches its greatest development in two separate groups of grass eaters; in the wombats the incisors in the upper jaw are also reduced to one and all the teeth are open rooted, and grow continuously through life as do the teeth of Rodentia and Lagomorpha. In the kangaroos the molar teeth are high crowned and the whole battery moves forward in the jaw, as in elephants, and the teeth are used successively as the more anterior ones are worn down and shed.

Foot structure

If foot structure is used as the criterion of grouping, the South American marsupials and the dasyurids in Australasia have in common hind feet with five separate subequal digits, whereas the peramelids and all the Diprotodonta have the first digit on the hind foot reduced to a nubbin and the second and third digits partly fused and together equal in size to the fifth digit (Fig. 1.5); the fourth digit is larger and in the hopping species is greatly elongated and larger than the lateral members. The semifused digits are used in grooming, but whether they have evolved primarily in response to this function or represent a progressive reduction of digits and more rigid foot for jumping locomotion can be disputed.

As with the teeth, the syndactylous condition may have evolved twice, for other characters set the Peramelidae well apart from the phalangerids and closer to the dasyurids.

Other anatomical features have been used from time to time to distinguish groups within the marsupials. The Diprotodonta possess a tract of nerve fibres, the fasciculus aberrans, linking the two cerebral hemispheres,[8] which is thought to be functionally analogous to the corpus callosum of eutherian mammals (see p. 205); it is not found in any Polyprotodonta, even the Peramelidae and Caenolestidae.

Urogenital tract

The female urogenital tract, so distinctive of marsupials, shows differences as between the several groups.[198] In the Didelphidae, Dasyuridae and Peramelidae, the urogenital sinus is long and the birth canal is also long and very transiently present at parturition; in the Phalangeridae both structures are shorter but the birth canal is transient, whereas in the Macropodidae it is usually permanent after the first parturition.

Spermatozoa

The morphology of the spermatozoa of marsupials can also disclose relationships.[38] In all species of American marsupials examined, some sperm in the epididymides of males or the vaginae of females after copulation are to be found in conjugated pairs. All pairs are united by juxtaposition of the shorter portions of the two heads. Sperm have not been found conjugated in the testis so it appears that the linkage occurs in the epididymis, and must be a quite firm bond to survive passage through the male genital ducts and deposition in the female. Conjugation of sperm is not known in any eutherian mammal nor has it been observed in any Australasian marsupial, so this character sets the two geographically separate groups of marsupials apart.

Within each family the sperm head has a characteristic shape, which distinguishes the Didelphinae from the Microbiotherinae and both from the Caenolestidae. In Australian species the shape of the head also distinguishes the Macropodidae, the Dasyuridae and the Peramelidae.[135] The wombat, *Vombatus ursinus*, and koala, *Phascolarctos cinereus*, have similar, hooked, sperm heads distinct from any other marsupials. The koala has been allied with the ringtail possum, *Pseudocheirus peregrinus*, on molar pattern but the sperm head of the latter species is very different from the koala's. As we shall see, other characters support the relationships of these three species, as disclosed by sperm morphology.

Other aspects of reproduction distinguish the several groups of marsupials but those will be considered in Chapter 2. On anatomical criteria the families of living marsupials separate into six distinct groups, which for convenience will be considered in the subsequent discussion of other criteria. The groups are:

Polyprotodonta

	Didactyla	1. Didelphidae
		2. Dasyuridae
		Thylacinidae
		Notoryctidae
		3. Caenolestidae
	Syndactyla	4. Peramelidae
Diprotodonta		
	Syndactyla	5. Tarsipedidae
		6. Phascolarctidae
		Vombatidae
		Phalangeridae
		Petauridae
		Burramyidae
		Macropodidae

Chromosomes

Karyotype

In recent years the chromosomes have been studied in a wide variety of marsupials,[176, 233] and shed more light on the inter-relationships of the families. The commonest number of chromosomes, or karyotype, possessed by eutherian mammals varies from 17 to 78 but is distributed around a single mode of 48. This has led to the conclusion that 48 is the ancestral karyotype and the higher and lower numbers have been derived from this respectively by division of two armed, meta-centric chromosomes to one armed acrocentric chromosomes, or by fusion of acrocentrics. On this hypothesis, the important number would be the number of arms rather than the number of chromosomes; it has been termed the 'nombre fondamental'.

Marsupials display a different pattern, which may help to elucidate these ideas. Firstly, marsupial karyotypes are much lower than the eutherian ones and are distributed bimodally around 14 and 22, with a range from 10 to 28; no species possesses the mean number of 18 (Fig. 1.6). The species with low numbers have a majority of metacentric chromosomes and the species with higher numbers have more acro-centrics, which tends to support the idea of the nombre fondamental although not unequivocally.

Family	Karyotype			
	< 14	14	16	20–28
Microbiotherinae		1		
Didelphinae		1		4
Caenolestidae		2		
Dasyuridae		10		
Peramelidae		5		1
Tarsipedidae				1
Burramyidae		5		
Petauridae				1
Phalangeridae				4
Phascolarctidae			1	
Vombatidae		2		
Macropodidae	2		10	14
Total Species	2	26	11	25

Fig. 1.6 Species of marsupials grouped according to chromosome number and family. The bimodal distribution is evident, as also the preponderance of 14 chromosomes in less advanced families. (Data from Sharman,[233] Martin and Hayman,[176] Gunson, Sharman and Thomson,[111] and Martin, personal communication 1970)

Furthermore, it has not been possible to decide whether 14 or 22 is the ancestral number, since species with 14 and with 22 chromosomes occur in most families. The recent observation that *Burramys parvus* has 14, like the other small Burramyidae,[111] and that the South American Caenolestidae and one species of the Microbiotherinae also have 14 chromosomes, favours 14 as the ancestral karyotype. Possibly as a consequence of their low chromosome numbers, marsupial chromosomes are considerably larger than those of eutherians. This fact has facilitated two avenues of research; identification of particular chromosomes is easier in marsupials than in Eutheria so that comparison of homologous arms can be made precisely, and also the activity of individual chromosomes of cells grown in culture can be done more readily than with eutherian cell lines.

Taking the first point, leucocytes from live animals are cultured in media which induces cell division.[176] The divisions are arrested at metaphase with colchicine, and the chromosomes swollen by changing the medium to a more hypotonic one. In the large clear chromosome figures thus obtained the length of the arms can be measured. To make meaningful comparisons between species, the relative DNA content of the nuclei of both species is measured and compared and then the proportion of the chromosome arms related to the DNA content. Assuming that the DNA is uniformly distributed along all the chromosomes, this proportional method enables direct comparison between related species. If two species share two or more chromosomes with both arms of equal proportional length the probability of them being related is very high.

Within the Dasyuridae there is very considerable homogeneity both of whole chromosomes and of their separate arms so that a 'standard dasyurid' karyotype can be described. Similarly the two species of wombat show close similarity as do 3 species of bandicoot, and the several kangaroos and wallabies with 16 chromosomes.

Comparisons between families show the sort of dissimilarities that might be expected from the knowledge of other criteria; thus the Dasyuridae and Peramelidae, with 14 chromosomes, do not have any chromosomes alike, nor do either of these families have chromosomes like those of the Burramyidae with 14 chromosomes. However, in *Cercartetus nanus*, the pigmy possum, (Burramyidae), 4 chromosomes are the same relative length as 4 dasyurid chromosomes, although the individual arms are different lengths. It is not possible at present to pursue this kind of analysis much further because the number of species examined in each family is still too small. Indeed, the examination of a third peramelid, *Macrotis lagotis*,[176] disclosed a karyotype of 20 whereas the three previous species examined have 14.

This still leaves undetermined the problem of the bimodal distribution of karyotypes. If 14 was the ancestral figure it is clear already from the newer analyses that the shift to higher numbers has occurred in at least four groups of marsupials independently. This again illustrates the point made in the context of dentition, that genetic changes may occur independently and yet appear superficially similar. The bimodal distribution of karyotypes in 4 separate groups of marsupials from both Australia and South America suggests that there is some fundamental selective advantage in arranging the chromatin into 14 or 22 chromosomes rather than into any other combination; it has been suggested that this may be related to the way in which the spindle forms at mitosis, a problem of fundamental interest to cell biology and one in which marsupial material may prove especially suitable for its elucidation.

Sex chromosomes

Four species of marsupial have unusual sex chromosomes, unlike the majority in which the female is XX and the male XY. The several patterns are set out below.

	♀	♂
Wallabia bicolor (swamp wallaby)[233]	XX	$/XY_1Y_2 + 8A$
Potorous tridactylus (rat kangaroo)[116]	XX	$/XY_1Y_2 + 10A$
Macrotis lagotis (rabbit-eared bandicoot)[176]	XX	$/XY_1Y_2 + 16A$
Lagorchestes conspicillatus (spectacled hare wallaby)[175]	$X_1X_1X_2X_2$	$/X_1X_2Y + 12A$

In the first 3 systems at meiosis the X and the two Y chromosomes form a trivalent from which it is interpreted[239] (Fig. 1.7) that the system is derived from a normal XX/XY system by fusion of the X chromosome to one autosome. Then the remaining autosome becomes the unusually large Y_2 and the original small Y becomes Y_1. This hypothesis has been tested very nicely by Hayman & Martin[116] using the culture technique described previously.

In some female mammals there is evidence that one of the X chromosomes is inactivated at an early stage of development, one aspect of which is a delayed synthesis of DNA by this chromosome at mitosis. If cells from a female are cultured in thymidine, labelled with radioactive tritium, only one X chromosome shows uptake of radioactivity. Now, if the fusion hypothesis above is correct, then in cells of a female rat kangaroo or swamp wallaby, only one arm of one X chromosome should show concentrated radioactivity while the same arm on the other X chromosome should be without radioactivity; the other arms of both chromosomes, representing original autosomes, should show moderate activity, like normal autosomes. This was precisely what developed in both *Potorous tridactylus* and *Wallabia bicolor*.

In *Lagorchestes conspicillatus* a similar fusion has occurred but between the original Y chromosome and an autosome so that the Y

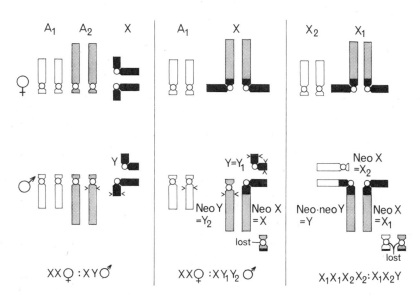

Fig. 1.7 Three patterns of sex chromosomes found in marsupials and a possible derivation. (Left) The chromosomes of a marsupial with an XX ♀ −XY ♂ sex-determining mechanism and two pairs of autosomes (A_1, A_2). Further autosomal pairs not shown. (Centre) A reciprocal translocation between one autosome and the X chromosome converts the system to an XX ♀ −XY$_1$Y$_2$ ♂ sex-determining mechanism and reduces the number of pairs of autosomes by one. (Right) Further reciprocal translocations between Y_1 and Y_2 chromosomes of the XY$_1$Y$_2$ system and between Y_1 and another autosome convert the system to an $X_1X_1X_2X_2$♀ − X_1X_2Y ♂ system reducing the number of autosomes by a further pair. ✕, Points of chromosome breakage and reunion; open circles, centromere regions of chromosomes. (From Sharman[239])

chromosome is unusually large and metacentric and at meiosis it pairs with the small X_1 and larger X_2, the latter being the original autosome. Again tracer studies identify the X_1 chromosomes as the true sex chromosomes. In this last example other indirect evidence supports the idea that there has been chromosome fusion; its nearest relative on other criteria, *Lagorchestes hirsutus*, has 20 autosomes and XX/XY sex chromosomes like all other small wallabies.

As in Eutheria the Y chromosome in marsupials is probably the male determining chromosome in sexual differentiation but this statement is based on examination of the karyotypes of only 6 intersexual marsupials. For two reasons these are very rare specimens; intersexes only come to the attention of interested people when large numbers of animals are being examined, and then the animals are usually dead before the anatomical anomalies are recognized, so that chromosome studies are impossible. However, chromosomes of three intersexual tammars, *Macropus eugenii*, one euro, *Macropus robustus*, and one brush possum, *Trichosurus vulpecula*, have been examined[246] (Fig. 1.8). A tammar lacking the Y chromosome and having the chromosome constitution XO + 14 was phenotypically female whereas the other, which had XXY + 14, was an abnormal male with undescended testes and a well-developed pouch and mammary teats. The third tammar, the euro and the brush possum each had normal XY sex chromosomes but were abnormally undeveloped males with undescended testes.

The small number and large size of marsupial chromosomes, especially those of *Potorous tridactylus* and *Wallabia bicolor*, has led to the use of marsupial cell lines in tissue culture studies. Cells from testis and kidney of *Potorous* have been grown through many generations in culture media,[285] and more recently, cells derived from the dasyurid, *Antechinus swainsonii*, with 14 chromosomes, have also been grown as permanent cell lines.[284]

Serology

The proteins synthesized by an animal are uniquely its own, whether they are formed into anatomic structures or are carried in its fluids, such as blood and milk. Some are sufficiently like those of other closely related animals that they do not elicit an antibody reaction when challenged in that species. The degree to which a whole suite of proteins in one animal react with those of another, immunologically, can therefore give a measure of relationship. As with other characters discussed earlier, confusion can enter due to convergence between less closely related stocks. However, the analysis of serum or milk proteins, because of the large number of characters involved, can provide more precision and a greater range of gradations in determining relationships than other criteria.

Two techniques have been used for separating and identifying proteins. A serum sample can be placed in a neutral matrix such as paper, starch or acrilamide gel, and an electric current applied. The size, shape and charge of the several proteins will determine how far each one will move in the matrix and separation of different proteins is thereby achieved. By appropriate staining, comparisons can be made of the serum

Species	Karyotype	Gonad	Urogenital tract	External appearance
1. *Macropus eugenii*	XO+14	ovotestes, abdominal	female	female phenotype, pouch with 2 small teats
2. *M. eugenii*	XXY+14	non-functional testes, abdominal	male	female phenotype, pouch with 4 teats, well developed penis
3. *M. eugenii*	XY+14	non-functional testes, abdominal	male	female phenotype, no pouch, empty scrotum, penis
4. *Macropus robustus*	XY+14	non-functional testes, abdominal	male	male phenotype, pouch with 4 teats, no scrotum
5. *Trichosurus vulpecula*	XY+18	non-functional testes, abdominal	male	pouch with small teats, no scrotum

Fig. 1.8 Intersexual marsupials. (Four specimens from Sharman *et al.*,[246] specimen 3 from own collection)

protein patterns from different animals. The other method depends upon the immunological properties of different proteins. If foreign serum be injected into a rabbit or other animal it will produce, after some days, antibodies specific for each protein in the injected serum. By challenging the rabbit serum with some of the original foreign serum, or antigen, a precipitate will develop. If serum from a closely related species be used to challenge the rabbit, a weaker precipitate will arise, the degree of precipitation being inversely proportional to the number of proteins the two species have in common. This can be refined by placing the rabbit antiserum in a central well cut in a block of starch and placing sera from a series of species related to the antigen donor in concentric wells in the same block of starch. The rabbit antisera and the several foreign sera diffuse slowly through the starch and when they meet precipitation will occur between the antiserum and the proteins held in

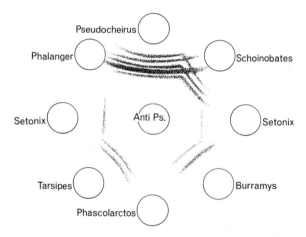

Fig. 1.9 The gel diffusion technique of Ouchterlony, used to determine serum affinities among several marsupials. Serum from a brush possum, *Trichosurus*, containing antibodies to serum from a ringtail possum, *Pseudocheirus*, is placed in the centre well. It has been prepared by injecting ringtail serum into the brush possum, so that antibodies are produced to all those proteins of the ringtail that are different from the brush possum's.

Normal sera from other species are placed in the surrounding wells and all diffuse through the gel centrifugally. Specific ringtail proteins are precipitated when they meet the anti-Ps serum from the centre well. The greatest number of precipitation bands occur between closely related species which share many proteins in common. In this example the greater glider, *Schoinobates*, is most closely related to *Pseudocheirus*, and the koala, *Phascolarctos*, the least. (From unpublished thesis of J. A. W. Kirsch, Comparative Serology of Marsupials, University of Western Australia (1967), by permission) the antigen. Since different proteins diffuse at different ~ecipitation bands may arise parallel to one another.

from Kirsch's work on marsupial serology,[147] antibodies ~~~~ in the brush possum, *Trichosurus vulpecula*, to serum from the ringtail possum, *Pseudocheirus peregrinus*, show a massive reaction to ringtail serum, as would be expected, and a reaction of almost equal magnitude to the serum of the greater glider, *Schoinobates volans*, but minor or no reactions to the sera of other marsupials. This result thus supports other evidence that the ringtail and greater glider are closely related to each other.

These techniques can be combined in several ways and by using different species as the source of antibodies independent checks can be made and refinements of the inter-relationships of species achieved.

From his extensive studies of the sera of 1800 specimens, Kirsch[147] concludes that as a group marsupials are more like one another serologically than they are like any eutherian, but that they have more affinities with these mammals than they have with monotremes. This accords with most current views. Within the marsupials, the American species studied are distinctly different from all Australian species. Figure 1.10 represents the affinities of all the species Kirsch has examined, including his recent unpublished results from South America.

Among the American marsupials the Caenolestidae are as distinct from the Didelphidae as both are from the Australian species. Within the Didelphidae, *Caluromys* and *Dromiciops* are distinct from the members of the sub-family Didelphinae, which agrees with their separation on anatomical grounds as the sub-family Microbiotherinae.

Within the Diprotodonta the affinities of the wombats, *Vombatus ursinus*, *Lasiorhinus latifrons*, and koala, *Phascolarctos cinereus*, are close and distinct from the ringtail *Pseudocheirus peregrinus*, and gliders, *Petaurus*, with which the koala has been sometimes associated on dental criteria. The three genera of small phalangerids with 14 chromosomes, *Burramys*, *Cercartetus*, and *Acrobates*, also have close affinities in their serum proteins. The Macropodidae separate, as on teeth, into the rat kangaroos, or sub-family Potoroinae, and the sub-family Macropodinae, which are a very uniform group, serologically. The distinctiveness of *Potorous tridactylus* and *Wallabia bicolor* on chromosome pattern is not upheld by an equivalent serum uniqueness within their respective sub-families, which further supports the hypothesis of recent chromosome fusion in these two species. Furthermore, the various kangaroo species with 16 chromosomes and the euro with 20 are serologically close. This is interesting because euros, *Macropus robustus*, will hybridize with red kangaroos, *Megaleia rufa*, although the offspring, like mules, are infertile.

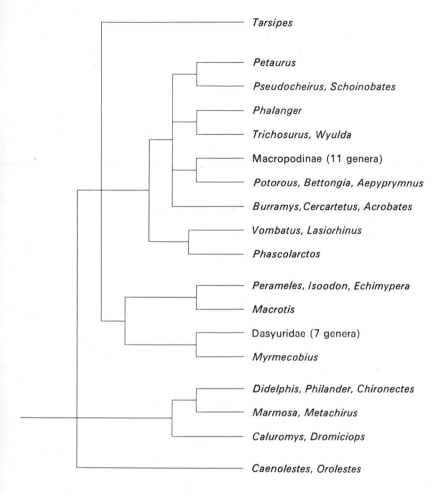

Fig. 1.10 Phenogram showing the serological affinities of 48 genera of marsupials. The linkages represent groupings of increasingly similar taxa from left to right. The abscissa is roughly linear, but the ordinate has no dimension. (Australian marsupials from Kirsch,[147] American marsupials by personal communication of Dr. Kirsch 1970)

At the species level serology has been used to elucidate the relationships between the grey kangaroos, *Macropus giganteus* and *Macropus fuliginosus*, throughout Australia.[148] These animals occur in all the temperate parts of the Australian Continent and in Tasmania and Kangaroo Island. At least 6 species have been described from different parts of Australia, but serological examination, as well as reproductive physiology and interbreeding experiments, have shown that there are only two true species, one extending from Queensland to Tasmania, and the other from Western Australia, through Kangaroo Island, to Victoria (Fig. 3.13). There is a zone of overlap in Victoria and southern New South Wales, where both species are sympatric, but do not naturally hybridize, although this can occur in captivity.

Another approach to intraspecific relationships is to relate serum analysis to the frequencies of the several genes controlling well defined blood proteins, such as transferrins and haptoglobins. This requires a breeding colony so that data on related animals can be compared, but in this regard marsupials are especially convenient because blood can be sampled from females and their pouch young at the same time. One such study has been done on the brush possum, *Trichosurus vulpecula*, which indicated differences in the distribution and frequency of certain genes within the separate populations of southern Australia (Fig. 4.1). These serum characters could be used as genetic markers in experimental studies by appropriate selection of stock from different localities, where particular genes are absent or very common.

This review of marsupial relationships emphasizes the kind of problems that still await solution. By comparing the various criteria available now it becomes clear that one must hesitate to ascribe the terms primitive or advanced to the several groups; rather one must acknowledge that no living species is any less evolved than another and all have diverged to different degrees from their remote predecessors. It is also evident that several remarkable convergences must have occurred independently in different groups so that it would be unwise to draw conclusions about closeness of relationship. What does become evident, however, is the unity of species within groups, regardless of criteria used, which encourages the belief that these are natural groups; how the separate groups are related in time, however, must await a far better fossil record than is presently available. These natural groupings within marsupials have usually been given familial rank within a single order of marsupials. To emphasize the wide variety of marsupials Ride[215] has proposed that the natural families be grouped into three orders. A classification which reflects this and incorporates the serological evidence is given in an Appendix.

ORIGIN AND DISTRIBUTION OF MARSUPIALS

North American cretaceous marsupials

Theories on the past history of animals are very dependent on the distribution of active palaeontologists and the techniques they use. In the last decade American palaeontologists have developed a technique for screening fossiliferous rocks under water, which recovers the tiny parts of small animals. As a result, it is now clear that marsupials comprised a large proportion of the mammalian fauna of the Upper Cretaceous of North America. They are associated with the remains of primitive Eutheria and many species of Multituberculates. Since similar studies have not been made on other continents it cannot be said that marsupials were restricted to that continent in the Cretaceous, although none has been found in extensive studies in Mongolia or in Europe, using the older methods of recovery.

In the lower Cretaceous rocks of North America marsupial and eutherian distinctions are less clear and one group of fossils represented by *Pappotherium* (Fig. 1.11) have been thought to possess characters intermediate between the two.[65, 159] Furthermore the pappotheriids can be derived from Jurassic eupantotheres.

Since these fossils are often represented only by fragments of jaws or even isolated teeth, it is pertinent to ask how they are identified as belonging to one or the other group. The molar teeth, being the most durable parts of the skeleton, provide the main criteria, which are derived by comparison with the more unspecialized, living representatives of the groups. The main derivation based on the molar teeth is shown in Fig. 1.11; in the simple tribosphenic teeth of the Eupantotheres, the upper molars bear on the posterior heel or talonid of the lower molars. In *Pappotherium* the upper molar has an incipient outer shelf, which is prominent in the later marsupials, such as *Alphadon*, and carries several cusps additional to the original three. Contemporaneous Eutheria, such as *Cimolestes*, did not have a stylar shelf; instead two additional cusps, the metaconule and protoconule, occur between the original three. In subsequent, Tertiary eutherians, such as *Hyracotherium*, a hypocone has been added to the labial surface and the metaconid has been lost from the lower molar, so that the two sets of teeth have become 'squared up'.

If these criteria are valid for discriminating the marsupials from eutherians, the dichotomy took place after the Lower Cretaceous. Slaughter,[257] however, has described the teeth of a Lower Cretaceous mammal, *Clemensia*, with a well developed stylar shelf (Fig. 1.11), contemporary with *Pappotherium*. This seems to indicate that the dichotomy may have occurred in the Lower Cretaceous.

A number of separate evolutionary lines are recognized among the Cretaceous marsupials of North America but only one group persisted into the Tertiary (Fig. 1.12). The early South American didelphids may be derived from one of these groups, although direct links are not known; no fossil mammals are known from the Cretaceous of South America and yet by the Palaeocene a variety of families is already established.

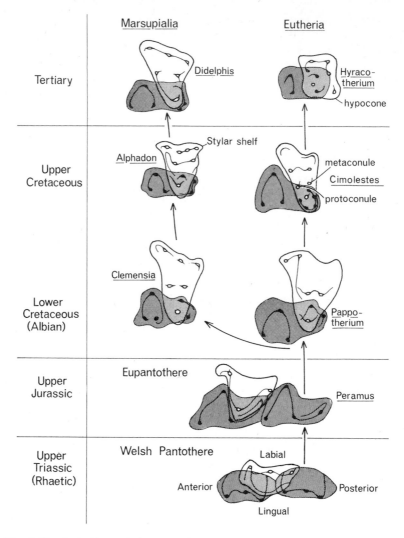

Fig. 1.11 Probable evolution of molar teeth in marsupials and eutherians from pantotheres. The lower left molar is shown in occlusal view, stippled, and the upper molar is superimposed in outline. (After Clemens,[65] Slaughter[257] and Romer[222])

The discovery of *Clemensia* opens the possibility for a much earlier spread of marsupials to South America but until Cretaceous fossils have been found this is conjectural.

South American tertiary marsupials

The Tertiary fossil record in both North and South America is well known. It is clear from this that the living marsupials are an impoverished remnant of a more abundant marsupial fauna,[197] which flourished successively in the Palaeocene and again in the Pliocene of South America (Fig. 1.12), whereas marsupials disappeared from North America entirely by the Miocene, as did the single line that reached Europe.

In the rich Palaeocene deposits of Patagonia the marsupial component of the fauna comprises the small carnivorous and insectivorous forms, as well as the first large carnivores derived from them, the Borhyaenidae. Herbivorous species with a reduced number of teeth and molars squared up for grinding, as in the Diprotodonta of Australia, also occurred in the Palaeocene but these Polydolopidae and Carolameghinidae became extinct by the Eocene, at the time that the first primitive eutherian herbivores appear in the fossil rocks. However, the carnivorous groups persisted through to the Pliocene and the Borhyaenidae flourished as the dominant carnivores on the continent. Some of them showed remarkable convergence to large carnivores of other continents: *Thylacosmilus* had very large upper canines resembling those of the eutherian sabre-tooth tiger *Smilodon*, and *Borhyaena* so closely paralleled in appearance the Australian marsupial *Thylacinus cynocephalus*, that it was considered at one time to indicate a close relationship of the two groups.

The Palaeocene Polydolopidae may have given rise to the Caenolestidae and possibly also to the Argyrolagidae[256] which were small jumping creatures resembling the Australian jumping dasyurid *Antechinomys spenceri*.

In the Pliocene or early Pleistocene, when the present land connection with North America arose, the long-isolated fauna of South America was profoundly disturbed.[197] Among the marsupials the Borhyaenidae and Argyrolagidae disappeared (Fig. 1.12) and the Caenolestidae became restricted to the Andean highlands. The Didelphidae, however, spread northwards through Central America and *Didelphis marsupialis* reached the Great Lakes (see Fig. 6.1).

Australian tertiary marsupials

Until the last decade very little was known of Australian marsupial antecedents, but recent diggings by parties from the University of California, the Australian Bureau of Mineral Resources and the South Australian Museum have disclosed fossil assemblages dating back some 25 million years to the early Miocene or possibly late Oligocene (Fig. 1.12). The Miocene deposits near Lake Eyre (see Fig. 1.13) contain representatives of most of the families still living and in some cases the

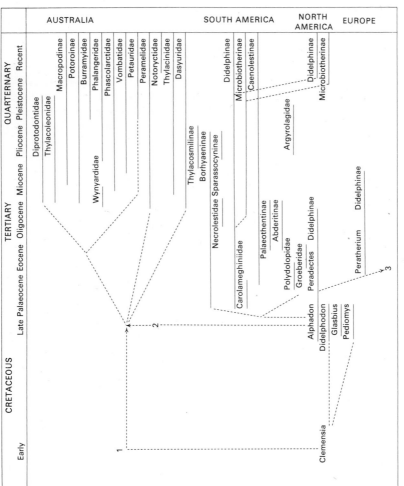

Fig. 1.12 Summary of living and extinct marsupial families. Known records shown by closed lines, and probable but unknown relationships shown by interrupted lines. Numbers indicate three possible routes for the origin of the Australian marsupials: 1. via Europe, Africa and Antarctica in early Cretaceous; 2. via South America and Antarctica in late Cretaceous; 3. via Asia in the

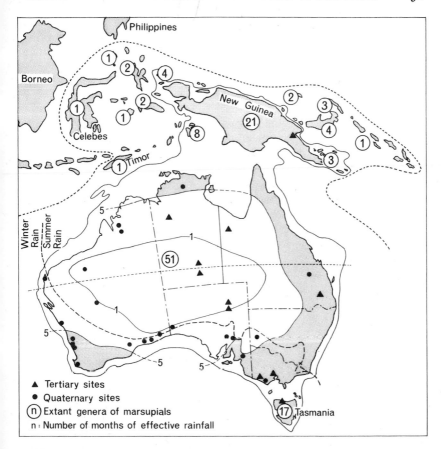

Fig. 1.13 The Australasian region showing the present climatic regions and limit of the continental shelf. Superimposed are the present extent (- - -), and distribution of marsupial genera and the sites at which Quarternary and Tertiary marsupial fossils have been found. (From Merrilees,[179] Ride,[6] Simpson,[255] Stirton, Tedford and Woodburne[262])

fossils can be ascribed to extant genera.[262] The only major groups represented in the Miocene by non-living descendants are the variety of very large marsupials of the family Diprotodontidae, Macropodidae and *Thylacoleo carnifex*, all of which became extinct in the late Pleistocene. Thus the great adaptive radiation of the marsupials must have occurred earlier than the Miocene and, judging by the degree of distinctness already achieved by then, probably in the Eocene or Palaeocene. One single fossil *Wynyardia bassiana*[215] is known at present from this whole span of time and it is late Oligocene; it is distinct from the later Phalangeridae but is thought to be antecedent to them, or to represent such a residual line.

The history of marsupials since the early Miocene is still sketchy. From the plant remains associated with the Lake Eyre sites (Fig. 1.13) and the occurrence also of fresh water species of turtle and crocodile it is surmised that the climate and vegetation was similar to much of present day New Guinea, warm and humid rain forest. In such an environment the large ground-living browsing animals presumably evolved from arboreal phalangers. The Diprotodontidae were a very successful group of marsupials if variety of species and wide distribution are the criteria of success. Three separate lines or sub-families have been distinguished and representatives of one or more are found in all the Miocene and Pliocene rocks. They were distinguished by large size and heavy build and *Diprotodon* itself bears a resemblance to an oversize wombat. Nevertheless, the teeth are unlike those of the wombat in having the anterior pair of cusps and the posterior pair of cusps in each molar united into transverse ridges.

Although rat kangaroos (sub-family Potoroinae) are represented in the lower Miocene rocks, true kangaroos (sub-family Macropodinae) do not appear until the upper Miocene.[262] It may be an accident of preservation that they have not been found earlier, but in subsequent beds they are common. These later rocks near Lake Eyre are associated with plant remains indicating a change in habitat to grasslands.

It has been suggested that the evolution of the Macropodinae, like that of the horses and ruminants in other continents, coincided with the appearance of grasses in the Miocene and the opportunity to tap this new source of food (see Chapter 3). Whether or not this is so, the Macropodidae and the Diprotodontidae dominated the grazing and browsing niches in Australia from the Miocene onwards. One group of kangaroos, the Sthenurinae, became large animals with short muzzles and some true kangaroos in the Pleistocene also reached giant proportions. The other large marsupial of Miocene and later times was the paradoxical *Thylacoleo carnifex*. This species is most closely related to the Phalangeridae and, although its teeth are reduced in number, the premolars are retained as very large shearing teeth. The Phalangeridae are generally not carnivorous but the skull of *Thylacoleo* gives the appearance of a powerful carnivore. Very recently the entire skeleton and associated skull of *Thylacoleo* has been discovered and the form of the clawed feet and powerful limbs support the view that this animal was indeed a large carnivore, notwithstanding its herbivore lineage. *Thylacinus cynocephalus*, now restricted to Tasmania, was also present from Miocene to Pleisto-

cene, when it and all the other large species became extinct on t
Australian mainland. Various ideas have been entertained for this
sudden extinction in the late Pleistocene and we will consider them
more fully in Chapter 7.

Origin of the Australian marsupials

Whereas the origin of the South American Palaeocene marsupials
from North America seems straightforward, the origin of the Australian
marsupials from this Cretaceous stock is another matter altogether.

Two mutually incompatible hypotheses have been proposed, neither
of which is really satisfactory. The northern hypothesis, most fully
expressed by Simpson,[255] proposes that early didelphid marsupials as
well as eutherians migrated from North America in the Cretaceous or
Palaeocene, via the Bering Straits to Asia and Europe, and that only small
rat-sized animals were able to filter through the Indonesian archipelago
towards Australia. By chance the only kind of mammals that successfully
reached the New Guinea-Australian land mass were small marsupials.
Being free of eutherian competition they radiated into all the main niches
as their cousins had begun to do in South America. Unlike in South
America, no eutherian followed and the marsupial radiation persisted
and flourished. This hypothesis requires the presence of marsupials in
Asia during the Palaeocene, but no marsupial has ever been found in any
Asian fossil bed, and in some Mongolian strata containing abundant
primitive Eutheria and Multituberculata, Marsupialia are conspicuously
absent. The best evidence at present is that marsupials, possibly related
to North American Eocene didelphids, occur in the Eocene of Europe,
and, since Europe and Asia were linked throughout the Tertiary,
marsupials could have been in Asia then. It still requires a formidable
exercise in logical gymnastics to conceive of small marsupials getting to
Australia, leaving no descendants on any intermediate island, while
equally small contemporaneous eutherians failed to get there but did
leave descendants on the northern islands. This hypothesis also presumes
that the Asian and Australian land masses had their present spatial
relationships at the beginning of the Tertiary, whereas much geological
evidence now supports the view that Australia was close to Antarctica
(Fig. 1.14) and has been drifting towards Asia in the Tertiary.[77, 259]

The idea that the continents were previously joined together was first
put forward 300 years ago and was championed more particularly by
Wegener in the early years of this century. But it was not regarded seriously by geologists until studies of paleomagnetism in ancient rocks from
different continents could only be reconciled by an alteration of the
relative position of the continents. Subsequently the fit of the continents

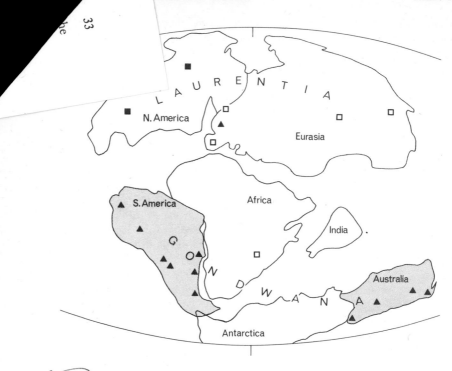

Fig. 1.14 Disposition of the continents at the beginning of the Cretaceous, according to Dietz and Holden.[77] Superimposed are the Cretaceous mammal sites that have yielded marsupials ▪, and those that have not ☐; also the main marsupial sites of the Tertiary ▲. (After Clemens,[65] Stirton et al.,[262] and Patterson and Pascaul[197])

to one another has been examined by computer analysis[259] and shown to be far closer than chance similarity would allow. Furthermore, the rocks from contiguous sides of a putative rift are found to match and the age of separation can be determined from the age of the oldest marine sediments overlying them. From these studies it appears that the southern continents of South America, Africa, Antarctica, India and Australasia were united as one super-continent of Gondwana until the end of the Jurassic, and that Eurasia and North America formed the second super-continent of Laurentia. During the Cretaceous these two super-continents fragmented and the component parts drifted away from each other: north and south America drifted westwards, Antarctica south, India and Australasia northward. India impinged upon Asia but Australasia has still some way to go. Thus the proximity of Australia and New Guinea to south east Asia is a relatively recent phenomenon and is thus an improbable route for the entry of marsupials in the early Tertiary.

The alternative hypothesis, expressed by Spencer and others at the turn of the century and revived recently by Ride,[216] is of a southern origin of the Australian marsupials from South America via Antarctica. This hypothesis gains support from the same geological evidence which indicates connections between these three continents in the Cretaceous. Western Antarctica is now thought to be a submerged archipelago, which in warmer climates might have acted as a species filter, such as Simpson postulates for the East Indies. On this hypothesis Cretaceous didelphids migrated to South America and some reached eastern Antarctica and thence Australia before the drift of continents had separated them too far apart. However, primitive Eutheria are associated with the earliest marsupials in South America and, if marsupials got across how did edentates and pyrotheres, condylarths and notoungulates fail to get through also? Cox[68] has proposed a temporal rather than a spatial filter to account for marsupials but not eutherians reaching Australia. His hypothesis rests on the idea, for which there is no fossil evidence, that marsupials antedated Eutheria and were able to reach Australia from North America via South America, Africa and Antarctica before the break up of Gondwana (see Fig. 1.14). The discovery of *Clemensia* indicates that marsupials might have been able to reach any continent joined to North America in the Early Cretaceous, but it does not allow for the exclusion of the contemporary pappotheriids.

Clearly the problems of the southern hypotheses are as intransigent as are those of the northern and will not be resolved until marsupial fossils are discovered either in Asia or in Antarctica; or until ancient eutherian fossils are discovered in Australia.

This latter possibility confounds a basic assumption with anthropomorphic undertones. Since marsupials flourished in Australia, the assumption runs, eutherians must have failed to reach Australia, because marsupials would inevitably lose ground to eutherian competition if these were present. If this assumption is set aside, a more parsimonious hypothesis may be proposed. It is that in the early Cretaceous, eutherians, multituberculates and marsupials all became equally distributed around the world, including Australia, probably by a southern route. With the subsequent separation of the continents, each carried its own cargo of primitive mammals, which then underwent adaptive radiation. It must be emphasized that at this time all marsupials and eutherians were small insectivorous animals and it is probable that neither organization was intrinsically superior to the other. The great eutherian and marsupial radiations on different continents in the early Tertiary, coincided with the radiation of the angiosperms as well as of the more advanced orders of insects. The mammals that first evolved adaptations to exploit these new food sources, and especially the exploitation of plant cellulose (see Chapter 3), may have gained such an advantage over all other kinds of mammals that the others disappeared. On each continent one or the other mammalian organization prevailed; in Europe and Asia eutherians prevailed over marsupials and multituberculates, in Australia the marsupials prevailed, while in South America both marsupials and eutherians prevailed over multituberculates until disturbed by the Pleistocene invasions from North America.

The subsequent course of evolution on each continent may have been determined, to some degree, by the size of the land mass and the variety of habitats available to the animals. Australia and South America were isolated throughout most of the Tertiary from the World continent, the parts of which were contiguous. Thus the mainstream of eutherian evolution took place on a land mass of about 90 million square km compared with the 17 million square km of South America and 7.8 million square km of Australasia. When these isolated faunas met successful representatives of the World continent fauna many succumbed, not because they were marsupials but because they were indigenes of smaller land masses.

This hypothesis, like its predecessors, has few facts to support it, especially the crucial evidence of early Tertiary fossils in Australia, but it makes less assumptions; it does not require spatial or temporal filters, nor does it assume a subjective and undefinable inferiority in the marsupial organization *per se*. This venerable assumption is not supported by the recent work on living marsupials, to be considered in the subsequent chapters.

2

Reproduction and Development

It will probably never be known whether the pantotheres had small, yolk-free eggs and brought forth young alive. However, the many differences between the Eutheria and Marsupialia in the adaptations for viviparity, suggest that true viviparity evolved independently in these two groups of mammals.

Viviparity confers selective advantages but also imposes severe consequences. These may be put in the form of 5 questions under which marsupial adaptations for viviparous reproduction and the development of homeostasis may be discussed.

1. How is synchrony of sexual development between male and female achieved to ensure internal fertilization?
2. How is the embryo maintained in the uterus?
3. How is synchrony between mother and embryo achieved?
4. How is the young succoured after birth until independence?
5. When does the young achieve homeostasis?

SYNCHRONY OF MALE AND FEMALE SEXUAL ACTIVITY

Very few species of marsupials are sexually active throughout the year. Most species have a fixed breeding season when the females are sexually active, but in some species the males as well show seasonal changes in the testes and accessory organs, correlated with sexual activity.

In eutherian mammals it is generally supposed that the breeding season occurs at a time of year that will ensure that the young are born in spring.

The bear is an exception in that the very immature young is born in winter but emergence from the lair occurs in spring. A similar relationship seems to hold for marsupials except that it is the emergence from the pouch rather than birth which occurs in spring. Thus species that suckle their young for a long time, such as the tammar, *Macropus eugenii*, breed in summer, whereas other species with a relatively short nursing period, such as *Antechinus stuartii*, breed in the winter. In places where spring is not a time of abundant food the indigenous marsupials, such as *Philander opossum* and *Didelphis marsupialis* in Nicaragua, do not show seasonal breeding. And again, in the desert regions of Australia where rainfall and feed are unpredictable the desert kangaroos (p. 127) and the desert-living dasyurids, *Dasyuroides byrnei*, and *Sminthopsis crassicaudata* (see p. 183), breed at all times of the year.

Both the males and females of the shrew-like *Antechinus stuartii* have a sharply defined breeding season in August to September.[293] In the males testicular growth and spermatogenesis begins in April-May and a month later the epidydimides have enlarged and sperm are beginning to appear in the urine. The prostate and cowper's glands, which provide the bulk of the seminal fluid, enlarge 100 fold in July at which time sexual activity begins. This is marked by increasingly aggressive behaviour towards other males and copulation with females. By September the testes have regressed markedly, the accessory glands are reduced in size and sexual activity has ceased. At about the same time the females are giving birth and thereafter suckle the litter in the pouch for three months. No other dasyurid has been studied so thoroughly as *Antechinus*, but it is known that in some of them also the breeding season is very restricted. Whether the male development is also highly synchronized remains to be discovered. In the Tasmanian devil, *Sarcophilus harrisii*, young are born May-June yet the testes of males examined in June lacked mature sperm, which suggests that in this species also testicular regression may occur very soon after the breeding season.

Sexual activity in the ringtail possum, *Pseudocheirus peregrinus*, and in the greater glider, *Schoinobates volans*, is also seasonal in both males and females though not as dramatically as in *Antechinus*. The testes of male gliders enlarge in January when spermatogenesis begins.[261] Spermatogenesis reaches a peak in May and then declines until it has ceased by July, when the size of the testes has again regressed. The enlargement is due mainly to growth of the seminiferous tubules; the Leydig cells of the interstitial tissue, which are the site of endocrine secretion, remain relatively constant throughout the year. Insufficient is known in this species about the accessory organs, such as the prostate, which might be affected by increased androgen secretion. However, the perianal glands which are present in both sexes of this species, are larger in males than females and secrete a pungent odour. The animals are well spaced out in the forests which they inhabit (see p. 166), living in the canopy of tall trees, and it is possible, though it has not yet been demonstrated, that the product of the perianal glands is used as a territorial marker.

In the brush tailed possum, *Trichosurus vulpecula*, and in *Antechinus*, males are often seen to rub the chest against various objects within their home territory. This sternal region has highly developed cutaneous glands, the oily secretion from which stains the fur an ochreous colour. The sternal glands are larger in the males than the females of these two species and become more prominent and active in the breeding season. In the male possum castrated before maturity the sternal glands do not develop but they will do so if the animal is injected with testosterone.[44]

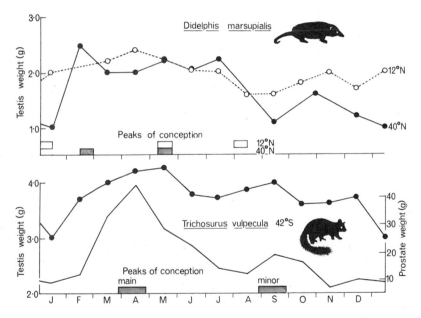

Fig. 2.1   Seasonal changes in testis weight of the opossum, correlated with latitude and the female breeding season. In the brush possum the prostate gland shows seasonal changes, not evident in the testes. (After Biggers[38] and Gilmore[98])

Although female brush possums have long been known to have a restricted breeding season, it had been thought until recently that the male was sexually active the year round. This was based on observations of testicular size and spermatogenesis which showed insignificant variation in different months of the year. The variation in the sternal gland, however, indicates changing endocrine activity and a careful study[98] of a large series of males taken throughout two years in New Zealand has disclosed that the prostate gland increases in size four fold between February and April (Fig. 2.1). Although the testis size does not change significantly during this period the amount of intersititial tissue in it does increase. It would seem from this evidence that in the possum, as in some Eutheria, pituitary stimulation of the interstitial tissue is independent of pituitary stimulation of the seminiferous tubules. In the economy of the brush possum the increased prostatic secretion at the time that the females come into oestrus may either facilitate frequent copulation or it may be important in effecting fertilization; female possums shortly after copulation contain 15–20 cm^3 of seminal fluid in the vaginal canals and culs de sac. Only during February to May could a male produce an ejaculate of this volume. These new observations are important because they demonstrate that it may be unwise to assume male fertility from testis size or spermatogenic activity alone.

The males of several species of Macropodidae, e.g. *Potorous tridactylus*, *Setonix brachyurus*, *Macropus eugenii*, in which the females are seasonally active sexually, are thought to be continuously fertile because active spermatogenesis occurs at all times of the year. However, in none of these species have the accessory glands been examined to determine whether there is a seasonal pattern of growth in these structures, as in the possum. In the tammar, *Macropus eugenii*, and in the grey kangaroo, *Macropus giganteus*, the vaginal apparatus is distended with seminal fluid after copulation to an even greater extent than in the possum so it would not be surprising if prostate size is shown to vary seasonally in these species too.

In the desert kangaroos, *Megaleia rufa* and *Macropus robustus*, which are facultative rather than seasonal breeders, there is no significant variation throughout the year in testis size, or in the volume of semen and density of spermatozoa in the ejaculate of adult males.[225] In comparison with eutherian mammals such as the bull and ram the volume of the kangaroo ejaculate is relatively greater whereas the density of sperm is much lower. The males considered on size and dental condition to be the oldest, showed reduced numbers of sperm but no diminution in the volume of the ejaculate. These then are the only two marsupials of which it can be confidently said that the males are continuously fertile and they are also the only two Australian species in which the females are certainly known to be continuously active sexually.

Two species of tropical marsupials from Nicaragua have been studied, one of which, *Philander opossum*, is restricted to the tropics. At the latitude of 12–13°N both species produce young at all times of the year.[38]

In males of *Philander* the testes show no seasonal variation in weight or spermatogenesis but in *Didelphis marsupialis* at this latitude[38] a few males showed reduced testicular size and activity in the second half of the year although the rest did not. In the same species collected in Pennsylvania (40°N) this pattern is much more marked (Fig. 2.1), the testes of all males caught from August to February being half the weight of those caught in the first half of the year, while in another study spermatogenesis was found to be incomplete in these months. In this region the

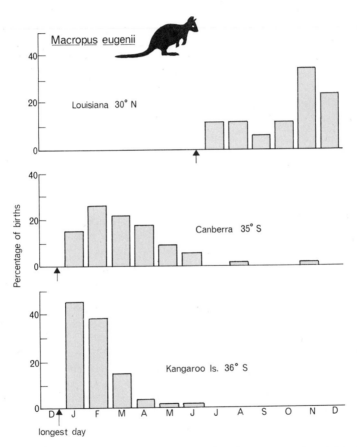

Fig. 2.2 Reversal of the breeding season in female tammars when taken from Australia to Louisiana in North America. In both places the first births occur a month after the longest day. (From unpublished thesis, P. Berger, 'The reproductive biology of the tammar wallaby, *Macropus eugenii*, Desmarest (Marsupialia)', 1970, Tulane University, by permission)

opossum only produces young in the first half of the year. Thus this species displays seasonal and non-seasonal breeding activity in different parts of its latitudinal range. The precise signal which determines the breeding season in this or any other species of marsupial is unknown. In higher latitudes when day length changes noticeably with season, it may be the proximate signal as it is in birds and mustelids. Thus opossums kept in Philadelphia by Farris[7] were induced to breed two months early in December by increasing the photoperiod in October to mimic the increasing day lengths which naturally occur after December 21.

Other species of marsupial with synchronized breeding seasons may be initiated into sexual activity by a decreasing photoperiod; most of the tammars on Kangaroo Island (36°S) and Bennett's wallabies, *Macropus rufogriseus*, on Tasmania give birth within a very short period of about two weeks at the end of January, just one month after the longest day. Tammars taken to Louisiana (30°N) by Dr. Patricia Berger reversed their breeding season and now give birth there at the end of June (Fig. 2.2).

Reproduction in male marsupials

Rete mirabile

Among eutherian mammals the optimum temperature for spermatogenesis is several degrees lower than deep body temperature; there is also evidence that low oxygen concentration (pO_2) is deleterious to spermatogenesis. The main adaptations are the scrotum and the pampiniform plexus developed betweeen the spermatic artery and vein. The scrotal position removes the testis from the region of deep body temperature to one which is usually cooler. If the ambient temperature rises above body temperature other devices such as evaporative cooling may be employed; if the ambient falls below the optimum the testes can be drawn nearer the body by reflex contraction of the cremaster muscle. The spermatic artery after it emerges from the inguinal canal is much coiled and is surrounded by the much branched spermatic veins to form the pampiniform plexus. The coiling of the artery has the effect of damping the pulse pressure while not reducing the mean blood pressure or pO_2, and heat exchange occurs between the arterial blood and the returning venous blood. In the dog and the ram the effect has been shown to cause a reduction of 5°C in the temperature of the testicular circulation as compared to the femoral artery.

Birds and monotremes retain the testes in the body and the former achieve spermatogenesis at higher temperatures than Eutheria, but marsupials carry the testes in a scrotum. In male *Trichosurus* F. Carrick has found that spermatogenesis is seriously impaired if the testes are experimentally returned to the body cavity, and similarly spermatogenesis was impaired in the few congenital cryptorchid intersexes referred to in Chapter 1[246] (p. 21). The marsupial scrotum is anterior to the

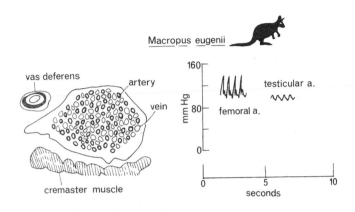

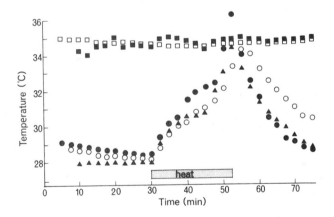

Fig. 2.3 Structure and functions of the rete mirabile of the male tammar, *Macropus eugenii*. (a) Cross section through the spermatic cord showing the vas deferens, the numerous divisions of the testicular artery and vein, and the cremaster muscle. (After Harrison[113].) (b) Pulse attenuation measured in testicular artery on surface of the testis. (c) Temperature in the rectum(□), testis(○), a testicular artery(●) and a vein(▲) on the surface of the testis and an internal spermatic vein at the inguinal ring(■) during heating of the testis of an anaesthetized tammar. (After Setchell & Waites[231])

penis and for this reason has a narrower stalk and is more pendulous than those of eutherians but the testes can be retracted by a cremaster muscle as in eutherian males. A countercurrent heat exchange is also found, although anatomically quite different from the eutherian. The spermatic artery divides into numerous branches which are intimately associated with the subdivided spermatic vein, thereby comprising a rete mirabile. The rete is variously developed in different marsupials.[27, 113] It is absent from the marsupial mole, *Notoryctes typhlops*, which carries the testes in the abdominal wall, and in small species, such as *Sminthopsis crassicaudata* and *Acrobates pygmaeus*, and is only slightly developed in *Phascogale tapoatafa* and *Potorous tridactylus*. It is best developed in the large members of each family, such as the tammar with 154 arterial branches (Fig. 2.3a) and the Tasmanian devil, *Sarcophilus harrisii*[110] with 15. This distribution suggests that its main function is for heat exchange, as in the eutherian pampiniform plexus, but some authors have argued that it may function as a venous pump and aid the return of fluid from the scrotum. This may be one of its functions, as it has recently been shown[231] that it damps out the pulse pressure without reducing the mean arterial pressure (Fig. 2.3b). However, its main function is undoubtedly to maintain a temperature differential of 4°C between the testis and the deep body temperature, as can be clearly seen in Fig. 2.3c for the tammar.[231] Similar results have been obtained in the red kangaroo, brush possum and Tasmanian devil.[110]

As well as these adaptations the tunica vaginalis, which surrounds the testis in the scrotum, is deeply pigmented in all American and many Australian marsupials examined. It has been suggested[38] that this may aid in heat loss by acting as a black body radiator under conditions of high ambient temperature but this interesting hypothesis needs further examination.

In this respect, as in so many others, the application of new techniques to marsupials reveals that they have independently achieved by slightly different means adaptations remarkably similar to those of their eutherian counterparts.

Prostate gland

Marsupials lack seminal vesicles but the prostate gland is more complex than that of eutherians and it has been surmised that the discrete parts are severally homologous to the separate accessory glands of other mammals. The marsupial prostate is a pear-shaped structure surrounding the urethra, from the epithelium of which it is derived in early development. Grossly and histologically it can be divided into anterior, middle and posterior regions.

In the opossum, *Didelphis marsupialis*,[172] the anterior region discharges granules by apocrine secretion from the apex of glandular cells. With the electron microscope these are seen to be lysosome vesicles and histochemically are shown to contain several enzymes such as acid phosphatase known to occur in mammalian semen. The epithelium of the middle portion consists of two types of cell, one synthesizing acid mucopolysaccharides and the other protein granules which may be enzymatic. The posterior region contains large amounts of glycogen which by enzymic action could give rise to fructose, a known substrate for spermatozoa. In Eutheria fructose is obtained from the circulation but it is possible that in the opossum it is produced by the prostate and that the enzymes from the other regions catalyse the hydrolysis.

Only one study has been made to discover the function of the marsupial prostate, using the brush possum, *Trichosurus vulpecula*.[131] In the intact possum about 30 cm^3 of semen containing 136×10^6 sperm per cm^3 can be obtained by electrostimulation of the anaesthetized animal. After resection of the prostate nothing was discharged except a small quantity of secretion from Cowper's glands which contained very few sperm. Nevertheless, the loss of the prostate did not impair spermatogenesis and motile sperm were delivered into the urethra and bladder. From this Howarth[131] concluded that the main function of the prostate is to provide a fluid vehicle for effective discharge of sperm. Although this may well be a very important function of the prostatic secretions it is likely that they also provide a metabolic substrate for the sperm and possibly other more subtle effects in the female as well.

Reproduction in female marsupials

Far more work has been done on female reproduction in marsupials than on male and generalisations can be made with more confidence. Detailed studies on reproduction of representatives from each of the extant families except the Caenolestidae have been made and in the Macropodidae many species are known.[7, 242, 278]

The female of a viviparous species has to resolve two conflicting functions of this mode of reproduction: to prepare the genital tract for the reception and conveyance of spermatozoa to the vicinity of ripe eggs, and conversely to prepare the tract to secrete egg coats and nourish the developing embryo. Some mammals have resolved the conflict by having a single and relatively long period of oestrus when the tract is receptive to spermatozoa, followed, either spontaneously or in response to copulation, by a progestational phase adapted to the requirements of the embryo. The main disadvantages of this pattern, as seen in the bitch or ferret, is that if the female fails to conceive during oestrus it remains infertile for the rest of that breeding season. A commoner, and therefore presumably, selectively more advantageous pattern is one in which periods of oestrus and progestation alternate cyclically, so that a female can conceive at a second or subsequent oestrus if the first is infertile. The acme of this polyoestrous pattern is seen in the murid rodents where the progestational phase is induced by copulation, while in the absence of copulation abbreviated cycles of recurrent oestrus occur, each one being potentially capable of developing to the next or luteal phase. This maximizes the chance of fertilization while also ensuring maximum support for the fertilized eggs.

The monoestrous pattern is seen in viviparous reptiles, probably in monotremes and in some of the dasyurid marsupials (eg. *Antechinus* and *Dasyurus*), but the great majority of marsupials are polyoestrous, like the majority of Eutheria. However, no marsupial is known to have evolved the short cycle of the Muridae and the oestrous cycles of marsupials vary in mean length from 22 to 42 days with the commonest length being 28 days.[7] These are mean lengths for the species and considerable variation occurs between different members of the same species and within one animal at different times of the year. Thus in female brush possums[203] isolated from males the cycles were a regular length of $25 \cdot 69 \pm 0 \cdot 31$ days during two periods of the year, February–April and June–August, while at other times oestrus occurred irregularly or not at all.

Ovarian changes in the oestrous cycle

The broad sequence of changes in the ovaries and genital tract are similar in monoestrous and polyestrous marsupials. The most complete information on a marsupial oestrous cycle is that of the brush possum, *Trichosurus vulpecula*,[203, 252] and what is known of other species seems to be similar. At the start of the breeding season the ovaries enlarge as a crop of follicles grow and mature. At oestrus a single Graafian follicle completes maturation in one ovary; ovulation occurs a day or so after oestrus, and the egg passes into the oviduct. A corpus luteum forms from the

---

Fig. 2.4 Oestrous cycle of the possum, to show the correlation between the changing structure of the cells of the corpora lutea and the level of progesterone in peripheral plasma and of relaxin activity in corpora lutea. The lipid vesicles probably contain cholesterol and subsequently pregneneolone which is converted to progesterone on the mitochondria at day 8–12, before being secreted into the adjacent capillaries. The small dense secretion, which appears later and is associated with the Golgi, is released between day 12 and 16, and may be relaxin.

The uterine gland cells begin to synthesize secretion by day 4 and maximum release occurs between day 8 and 12, after which the exhausted cells are replaced by an underlying epithelium, newly formed from stromal cells. (Drawings made from electronmicrographs provided by C. D. Shorey, progesterone data by permission of G. B. Thorburn,[252] relaxin data from Tyndale-Biscoe[279])

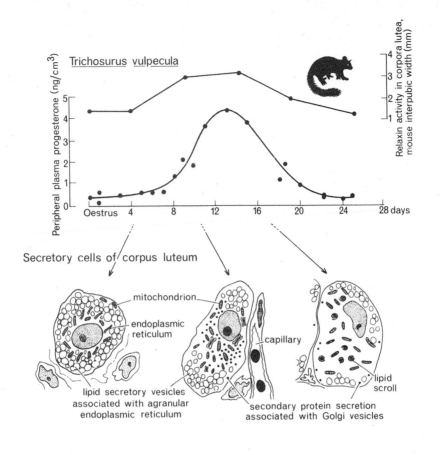

Trichosurus vulpecula

Peripheral plasma progesterone (ng/cm³)

Relaxin activity in corpora lutea, mouse interpubic width (mm)

Oestrus 4 8 12 16 20 24 28 days

Secretory cells of corpus luteum

mitochondrion

endoplasmic reticulum

capillary

lipid secretory vesicles associated with agranular endoplasmic reticulum

secondary protein secretion associated with Golgi vesicles

lipid scroll

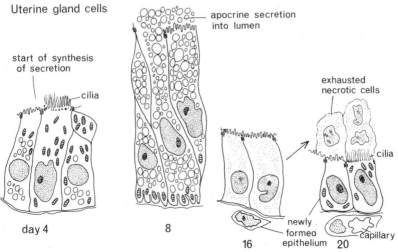

Uterine gland cells

apocrine secretion into lumen

start of synthesis of secretion

cilia

exhausted necrotic cells

cilia

day 4

8

16

newly formed epithelium

20

capillary

wall of the ruptured follicle, which consists of two layers of cells; in the formation of the corpus luteum the cells of the outer layer penetrate the inner layer, laying down a network of fine capillaries and lymphatic ducts on a connective tissue framework. The cells of the inner layer or membrana granulosa enlarge and differentiate into endocrine secretory cells, known as luteal cells. They provide the bulk of the tissue so that the gross dimensions of the corpus luteum are an approximate index of glandular activity. More precise evidence of glandular activity is now available from electron microscopic studies of the corpus luteum throughout the possum oestrous cycle, and correlated measurements of progesterone concentrations in the blood leaving the ovarian vein and in the peripheral circulation.

From the fourth day after oestrus the luteal cells are actively synthesizing a lipid-bound material which is enclosed in vesicles. These first appear around the central nucleus in association with well formed smooth endoplasmic reticulum, and they increase in number until, by day 8, they fill the cells (Fig. 2.4). At this stage another kind of secretory product makes its appearance within the Golgi vesicles around the nucleus, and the lipid vesicles are displaced to the periphery. They are small, electron-dense bodies, most probably containing a protein or polypeptide, which increase in abundance as the number of lipid-filled vesicles declines after day 12, and then themselves decline to insignificance by day 18. By this time the cells look degenerate with crumpled nuclei and shrunken cytoplasm. The gross size has declined also and in the unmated possum a new crop of follicles are beginning to enlarge as pro-oestrus supervenes.

The progesterone concentration in the peripheral circulation (Fig. 2.4) rises from about $0·5$ ng/cm^3 at oestrus to a peak of more than 4 ng/cm^3 at day 12; it then declines steadily to $0·5$ ng/cm^3 again by day 20. The pattern thus disclosed reflects the structural changes seen with the electron microscope and is strong evidence for the view that the lipid-bound material is progesterone or a steroid precursor of it. In the red kangaroo progesterone has been detected in the corpus luteum at concentrations of $0·5$ to $3·0$ μg per gland, and in the opossum, *Didelphis*, progesterone has been synthesized by slices of luteal tissue cultured in medium containing cholesterol marked with a radio-active carbon atom.[67] The evidence for progesterone secretion by marsupials has been hard to get because of the very low concentrations in which it occurs in the gland and in the circulation.

Conversely, the low blood concentrations may indicate a very rapid catabolism of progesterone in the liver and its excretion as an inactive metabolite. One such metabolite, pregnanediol, has been detected in the urine of female possums[203] at a concentration of 100 μg per day during days 10 to 16 but it was also detected at a concentration of 40 μg in males.

A correlation has also been detected between hormone activity and the pattern of synthesis of the other secretory vesicles observed in possum luteal cells between day 8 to 16. Suspensions of possum corpora lutea from days 9, 12 and 15 after oestrus showed a positive response in the assay for the polypeptide hormone relaxin,[279] but a negative response at oestrus and on days 4 and 19 (Fig. 2.4). In Eutheria relaxin specifically softens the connective tissues of the genital tract and the pubic symphysis during late pregnancy and the assay used for the possum ovaries was the degree of separation of the symphysis pubis of immature female mice. The correlation of positive activity in possum corpora lutea with the appearance of the small secretory vesicles is close and coincides with the time of the cycle when parturition would occur in a pregnant possum.

Changes in the vaginal complex in the oestrous cycle

The changes described in the ovary are reflected in changes in the different parts of the genital tract. The pro-oestrous phase, while the follicles are maturing, is associated with active secretion of mucopolysaccharide by the epithelial cells of the oviducts; by hyperplasia of the uterine glands and epithelia; and by hypertrophy of the vaginal complex and increase in thickness of the epithelia of this structure. This reaches its climax at oestrus when the lateral vaginae and the vaginal expansions may have enlarged several times from the quiescent size. One part of the complex enlarges more than the others. Thus in the kangaroos and wallabies and the American didelphids the lateral canals enlarge, whereas in the possum the median vagina enlarges and in the rat kangaroos and one hare wallaby a special outgrowth of the median vagina, the vaginal caecum becomes enlarged. This enlargement is accompanied by fluid secreted into the lumen from the glandular epithelium of the median vagina and by sloughing of cornified cells which provide the evidence of oestrus in a smear taken from the urogenital sinus.

In the monoestrous marsupials, such as *Antechinus*, an oestrous period lasting several days occurs once in the year, closely coincident with the period of maximum sexual development of the male, and copulation takes place several times during this single period.[171] Ovulation occurs sometime during or after oestrus, and is not closely tied to copulation. It is therefore not surprising to find that in this species, and in *Dasyurus viverrinus*,[125] sperm are retained alive in the fallopian tubes for many days after copulation and are thus available to fertilize the eggs when they are eventually shed. This contrasts with the polyoestrous species in which copulation is generally restricted to a brief period of less than a day, immediately prior to ovulation, and sperm do not survive in the uterus for much more than a day.

The great size of the vaginal canals in oestrous kangaroos led earlier workers to suppose that they were an adaptation for sperm storage and delayed fertilization, but this has been disproved as sperm do not remain for more than a few days in the female tract of macropodids. However, as mentioned earlier, after copulation the vaginal complex becomes grossly distended with seminal fluid, some of which may coagulate to form a plug in the urogenital sinus. It is by no means clear what the significance of this is, but the facts that in the possum the prostate glands of the males enlarge greatly at this time, as already mentioned, and that the vaginal apparatus of the females also enlarges considerably, do not seem like fortuitous adaptations without selective advantage. Prostatic secretions in other mammals contain prostaglandins and other substances which stimulate the myometrium and may thus aid the transport of spermatozoa up the uterus and fallopian tubes. What is known of the marsupial prostate indicates that this aspect of marsupial reproduction would repay further research.

Uterine changes in the oestrous cycle

After oestrus there is a diminution in the size of the vaginal complex but increasing activity in the uteri, which become more vascular and oedematous, while the gland cells become transformed to an active secretory condition. The uterine glands undergo hyperplasia during and after oestrus and, in the brush possum,[46] this appears to be greater in the uterus contiguous to the ovary bearing the developing follicle or new corpus luteum. Subsequently this uterus is larger than the other as the greater number of cells hypertrophy. It is not clear how this unilateral effect of the ovary is achieved, but it is probably via a vascular connection between the ovary and uterus, since it was shown by ligating the oviduct that the stimulus cannot come from follicular fluid as it passes down the oviduct after ovulation. Whatever the means, it is possible that it has an effect later in the cycle; in the tammar wallaby, another monovular species, flushings from the uterus associated with the new corpus luteum consistently contain greater amounts of protein than the opposite uterus, which again may reflect the presence of more secretory cells in the ipsilateral uterus than in the other.

As the corpus luteum grows the gland cells of the uterus are transformed from cuboidal cells with large central nuclei to elongate columnar cells with small basal nuclei.[203] This characteristic form has been termed the luteal phase because it is induced by the corpus luteum secretions or progesterone substitution. The luteal phase in the possum extends from day 4 (when progesterone begins to increase in the blood) to day 18.

With the electron microscope it is seen[252] that the region around the nucleus is filled with granular endoplasmic reticulum, interspersed with mitochondria, clear evidence of active synthesis (Fig. 2.4). On day 8 synthetic activity begins to decline and the accumulated secretion pours from the apices of the cells and may even include components of the cells as well. This material continues to flow into the lumen of the uterus for several days, reaching a peak at day 12 and declining thereafter. Uterine gland cells of the tammar in the luteal phase show a similar appearance under the electron microscope and in this species the secretion has been analysed.[211] Free amino acids have not been detected but carbohydrate and protein are present. By gel electrophoresis the protein is found to be composed almost entirely of albumin and pre-albumin fractions (Fig. 2.9), whereas the slower moving proteins, characteristic of blood serum, are not detectable. This indicates that uterine fluid is a product of active secretion rather than an exudate of lymph or serum, as the earlier workers believed.

By day 18 in the possum the gland and epithelial cells are exhausted and dying. Meanwhile cells from the stroma of the uterus, having assembled under the basement membrane, establish contact with each other by tight junctional complexes and produce a new basement membrane outside the old one. When completed, the old epithelium sloughs away and is phagocytosed by polymorph leucocytes, and the new layer proliferates a new epithelium in readiness for the next cycle. A similar regeneration of the uterine epithelium and glands probably occurs in the opossum, *Didelphis marsupialis*,[114] the rat kangaroo, *Bettongia gaimardi*[90] and the quokka, *Setonix brachyurus*,[274] although the authors of these earlier observations did not interpret them in this way because they did not have the greater resolving power of the electron microscope with which to unravel the process.

In the closing stages of the luteal phase of several species, and after, the sub-epithelial capillary bed becomes more prominent than hitherto and more closely applied to the epithelial basement membrane, so much so that some capillaries appear to bulge out beyond the margin of the epithelium. In actual fact the epithelium maintains its integrity but by displacement of cytoplasm the capillary is separated from the lumen only by the two plasma membranes of the epithelial cells and their common basement membrane. Thus two phases can be distinguished in the cycle; a secretory phase controlled by the corpus luteum, and a vascular-regenerative phase during the regression of the corpus luteum.

These conclusions are further confirmed by the results of surgical removal of the corpus luteum or both ovaries at different phases of the cycle.

In the possum and the quokka, both monovular species, removal of the corpus luteum during the first week after oestrus leads to premature ovulation from the remaining ovary and failure of the luteal phase to be sustained in the uteri;[203, 275] if the operation is performed during the luteal phase, however, it does not affect the subsequent maintenance of this phase in the uteri nor the time of the next ovulation. Thus the corpus luteum appears to inhibit follicular development during the first half of the cycle and it initiates but does not maintain the luteal phase in the uteri. The length of the oestrous cycle is thus determined in each species by the time it takes for follicles to mature and by the duration of the corpus luteum inhibition of this process.

The luteal phase can be induced in the uteri of anoestrous or ovariecto-mized quokkas by giving injections of progesterone. However, injections of relatively large doses to animals, deprived only of the corpus luteum, failed to arrest follicular growth, which suggests that some other product of the corpus luteum may be responsible for inhibiting follicle growth. In the red kangaroo, *Megaleia rufa*,[243] also, the corpus luteum has been shown to be capable of inhibiting follicular growth even when, during lactation, it is insufficiently active to induce a luteal condition in the uteri (see p. 54).

The corpus luteum has other as yet less defined effects on the female marsupial; in the possum the pseudovaginal strand of connective tissue through which the young passes at birth, becomes loose and oedematous at mid-cycle, a condition that can be induced in ovariectomized possums with progesterone. So far no direct effect of relaxin on the possum genital tract has been detected but the loosening of the connective tissue surrounding the uterus and the birth canal and the enlargement of the pouch are likely sites for its action.[277] The pouch lips become thicker and moister during the luteal phase and the mammary glands develop to the same degree as those of pregnant animals so that such animals, even virgin, are capable of suckling young transferred to their pouches at this time (see p. 88).[235]

The role of the pituitary in eutherian reproduction is well known but its role in marsupials is only now beginning to be studied. In Eutheria two moieties with gonadotrophic effects have been recognized but there is doubt as to whether they are always separate hormones or not. Follicle stimulating hormone (FSH) initiates follicle growth and maturation, while luteinizing hormone (LH) causes ovulation to occur and may be involved in the subsequent maintenance of the corpus luteum. These effects are known from experiments involving removal of the pituitary, hypophysectomy, by substitute injections and, more recently, by measuring the amount of gonadotrophins circulating in the blood of intact animals. J. P. Hearn[116A] has developed a method for measuring total gonadotrophin in the blood plasma of tammar wallabies and a technique for hypophysectomy, and he has allowed me to quote some of his results here and in a later section (p. 72). The basal level of gonado-trophin in both males and females was 2–5 ng/cm³ plasma throughout the year, except in females on the day of oestrus when the level was 11–18 ng/cm³. No gonadotrophin was found in hypophysectomized animals so that the activity measured was of pituitary origin. Removing the pituitary from females at mid-cycle prevented follicle growth and subsequent ovulation, but hypophysectomy performed after ovulation did not stop subsequent development of the corpus luteum or the uterus.

These results indicate that secretions of the pituitary are necessary for follicular growth and ovulation in the tammar but are not needed to maintain the remainder of the cycle. In earlier experiments on several species of marsupial[7] other workers had observed that FSH of eutherian origin induced follicle growth but not ovulation, while LH induced luteinisation and the synthesis of progesterone in cultured slices of opossum ovaries.[67]

HOW IS THE EMBRYO MAINTAINED IN THE UTERUS?

Relationship of gestation to the oestrous cycle

In all except one marsupial, conception does not affect the course of the oestrous cycle, provided the young are removed at birth and prevented from suckling. This is in marked contrast to most eutherians in which oestrous cycles are suppressed during gestation and, to a lesser extent, during suckling as well. There are, of course, some eutherians, such as the ferret and the bitch, in which oestrus is not suppressed by pregnancy because the length of the cycle is equal to the length of gestation, and they resemble the prevalent marsupial condition, in which gestation occupies a shorter time than one oestrous cycle. The time may be very much shorter as in the opossums, Didelphis marsupialis[213], Marmosa mitis and the bandicoot, Perameles nasuta,[134] or only a day or so less, as in most macropods; in only two marsupial species yet studied does gestation extend past one oestrous cycle and in only one of these species, the grey kangaroo, is oestrus suppressed, as it is in most Eutheria.

Young of different marsupials may thus be born at any phase of the cycle (Fig. 2.5) from the height of the luteal phase in Perameles through the post-luteal phase in the opossum and brush possum, to pro-oestrus and even post-oestrus phases in various Macropodidae. In the normal course of events the newborn young attaches to a teat and starts suckling, and this stimulus blocks the oestrous cycle at the stage reached at parturition; the cycle only resumes if suckling ceases, either artificially by removing the pouch young (RPY), or naturally when the young

54

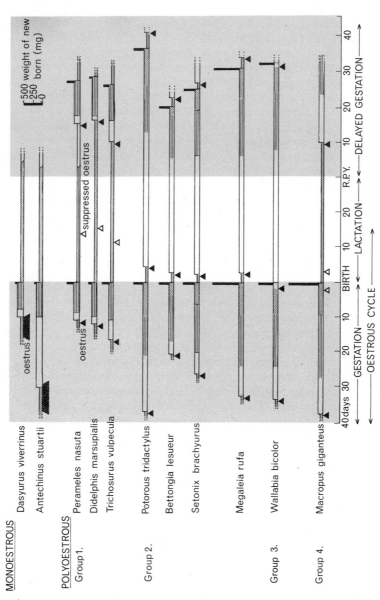

Fig. 2.5 The reproductive cycles (▲–▲), gestation lengths (▲–■) and neonatal weights of 11 marsupials compared by reference to the day of birth. Subsequent lactation suppresses the reproductive cycle at approximately the stage reached at parturition, and removal of the pouch young (RPY) releases it from inhibition, so that the suppressed events are completed. The successive stages of the cycle are represented in the horizontal columns as follows: preluteal, blank; luteal, horizontal hatch; post-luteal/pro-oestrous, stipple.

vacate the pouch. Thus in the bandicoot, large, apparently active corpora lutea persist throughout lactation and only decline if this stops. In the opossum and brush possum the corpora lutea continue to decline during lactation but subsequent follicles do not grow and ovulate; while in many macropod species, ovulation occurs soon after parturition but the growth of the corpus luteum and the luteal phase is suppressed during lactation and only resumes if this stops. Finally, in the grey kangaroo, *Macropus giganteus*,[64] gestation exceeds the length of the cycle so that oestrus is suppressed until the end of lactation or until lactation is artificially stopped by removing the pouch young.

From these comparisons it seems clear that the corpus luteum, which is a functional endocrine gland in the oestrous cycle, does not have an additional role in the maintenance of gestation, nor does parturition invariably coincide with its involution. Furthermore, since its life is not prolonged during gestation, it is unlikely that it is influenced by gonado-trophic hormones secreted by the embryo or placenta, as in many Eutheria. It is possible that the persistent corpora lutea of the bandi-coot[134] may be actively maintained by the pituitary during lactation but until critical experiments demonstrate this, it is simpler to see the phenomenon in the general context as a blockade of the oestrous cycle imposed by suckling, which in this species precedes the involution of the corpora lutea.

The role of the corpus luteum in gestation has been examined in four species. In three monovular species the effect of removing the corpus luteum could be distinguished from the effect of removing both ovaries, and the contained corpora lutea, entire. In the quokka[275] and in the brush possum[203] the embryo would continue to develop normally to full term if the corpus luteum was removed after the seventh day of gestation, but would die if the corpus luteum was removed somewhat earlier. The essential factor for gestation to continue was that the luteal phase be established by the time of operation; once established the secretion produced is evidently adequate to maintain development of the embryo. In these two species, and in the tammar, essentially the same results are obtained when both ovaries are removed after day 7.[280] Thus the factor that determines the duration of intra-uterine development is not the corpus luteum nor the other ovarian tissue but is more likely to be found in the embryo itself. Much earlier experiments by Carl Hartman[115] on the opossum were less clear cut; he removed both ovaries and con-cluded that the corpora lutea are essential for the full 13 day gestation of the opossum. His conclusions have been widely quoted since as representing the marsupial condition, but recent re-examinations of his results by Buchanan[55] show that in some of his animals the embryos clearly survived and continued to develop for several days after the operation. It would be interesting for someone to take a fresh look at opossum gestation in the light of the recent Australian work to see if there is a real difference in the role of the corpus luteum between these two groups of marsupials.

A long-standing but erroneous idea about marsupials is that they are characterized by very short gestation lengths compared to their eutherian counterparts. This derives, again, from early observations on *Dasyurus viverrinus* and *Didelphis marsupialis*. There is no doubt now that these two species have short gestation periods of 9 and 13 days respectively but recent observations on many other marsupials show that this is not so in all species. Thus[7] in *Antechinus* the gestation period of 25–31 days is longer than that of a comparable sized eutherian shrew, *Blarina brevicauda*, while the rat kangaroos have gestation lengths slightly shorter or longer (Fig. 2.5) than equivalent eutherian herbivores such as the rabbit. Only the larger macropods have gestation lengths substantially shorter than equivalent-sized eutherians such as antelopes.

The real point at issue is not gestation length, which is variable, but the biomass of young produced at the end of gestation. On this count all eutherian mammals bring forth very much larger young at a much

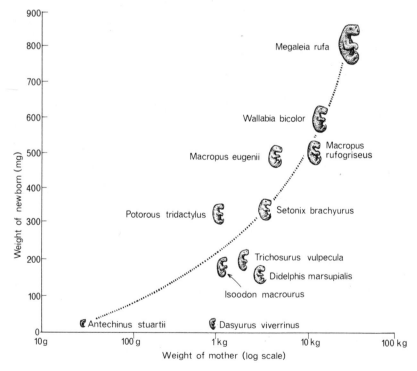

Fig. 2.6 Relationship between log of maternal weight and neonatal weight and development. Young are drawn to scale × ⅓ life size.

more advanced stage of development than do any marsupials. Whereas Eutheria include growth phases as well as embryogenesis in the uterus only embryogenesis occurs in the uterus of marsupials and the main growth occurs in the pouch.

Notwithstanding their wide range of adult size, newborn young of all marsupials are remarkably uniform in size (Fig. 2.5), ranging from 10 mg to 750 mg. Since those with the smallest offspring bring forth litters of 20 at a time, the biomass at birth is much closer than this range implies. There is a correlation within marsupials between birth weight and the log of maternal weight (Fig. 2.6) and there are slight differences in the developmental stage of the young at birth but these are trivial compared to the uniformity which prevails despite differences in total gestation length.

These several observations lead to two conclusions. One is that the marsupial placenta is a much less efficient transport system than the eutherian, and we will consider this later. The second is that it functions for a much briefer time within the total gestation period. If the latter, then a large proportion of marsupial gestation is taken up by the free, unattached, phase of development. We will now consider the development of the embryo in this context.

Intra-uterine development

By eutherian standards marsupial eggs are large, although very much smaller than the yolky eggs of monotremes and reptiles. The eggs of some marsupials (eg. *Dasyurus*) contain a yolky material, which is extruded by the fertilized egg and becomes enclosed within the cavity of the developing blastocyst (Fig. 2.7a). Its true nature and function are still unknown but it has been homologized with the non-cellular yolky centre of the monotreme egg and is used as evidence for a closer relationship with these mammals. It seems unlikely, however, that it would have persisted for more than 100 million years without having as well some present value in the economy of its possessor.

The egg is shed from the ripe follicle at ovulation, having extruded the first polar body, and is surrounded by a thin membrane, the zona pellucida, derived from secretions of the follicle cells. It is not, however, surrounded by a cloud of these little cells as is the eutherian egg. It passes into the open funnel of the oviduct and is carried rapidly along this to the uterus, which it reaches in about a day, unlike eutherian eggs which take several days to cover this distance. During this short time it meets the spermatozoa and a single one penetrates the zona pellucida, stimulates extrusion of the second polar body and effects fertilization. Many other spermatozoa become embedded in a second membrane laid down around the zona and varying in thickness from $1 \cdot 5 \, \mu$ to $8 \cdot 0 \, \mu$ (Fig. 2.7a). By analogy with birds this layer was called an

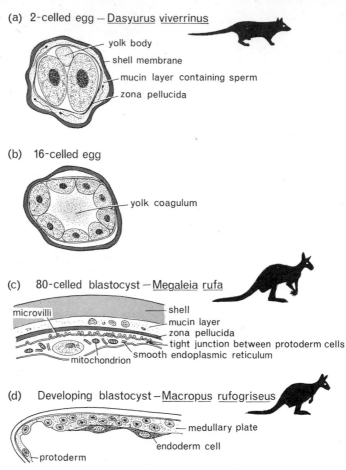

(a) 2-celled egg – <u>Dasyurus viverrinus</u>

- yolk body
- shell membrane
- mucin layer containing sperm
- zona pellucida

(b) 16-celled egg

- yolk coagulum

(c) 80-celled blastocyst – <u>Megaleia rufa</u>

microvilli
- shell
- mucin layer
- zona pellucida
- tight junction between protoderm cells
- smooth endoplasmic reticulum
- mitochondrion

(d) Developing blastocyst – <u>Macropus rufogriseus</u>

- medullary plate
- endoderm cell
- protoderm

Fig. 2.7 Representative stages of early development in three species. (a) Egg after first cleavage and extrusion of the non-cellular yolk body; three membranes distinct. (b) Formation of the blastocyst after the fourth cleavage and liquefaction of the yolk body (From Hill[122]). (c) Drawing from electronmicrograph to show detail of undifferentiated protoderm cell. (d) Similar blastocyst after formation of endoderm cells and differentiation of medullary plate, destined to become the neural tube of the embryo. (After Hill[122])

albumin layer by the earlier writers, but recent histochemical and electron microscopic work by C. D. Shorey and R. L. Hughes has shown that it is an acid mucopolysaccharide secreted by glandular cells of the oviduct. It is homologous with the mucin coat of rabbit eggs, but the marsupial egg receives as well a third coat not possessed by any eutherian. It is a keratinous shell membrane, secreted by cells of the lower segment of the oviduct and the uterus and is thought to be homologous with the basal layer of the leathery shell of monotreme and reptile eggs. The shell membrane has remarkable properties, for it is resistant to enzyme digestion, is inelastic and yet is capable of extreme attenuation. It surrounds the embryo throughout the short gestation of such species as the brush possum and through most of the longer gestation of macropodids such as the quokka.

The early development of the marsupial zygote excited interest many years ago because of its resemblances to the early development of monotremes and this interest has been revived recently by palaeontologists[159] concerned with the relationships of Marsupialia to the Pantotheria and Eutheria. So far, however, the renewed interest has not led to a full re-examination of marsupial development with the electron microscope or the biochemical techniques being used on other embryos, so we must rely on the older studies of Hill,[122] Hartman[3] and McCrady[164] and pose the questions that await solution.

After the yolk body has been extruded, the fertilized egg cleaves meridionally three times to form eight blastomeres which form a ring around the yolk body. Each of these cells now cleaves equatorially so that two rings of eight cells result. Further meridional and equatorial divisions result in the formation of a hollow sphere of 60 to 80 cells arranged in a single layer (Fig. 2.7b). This is the unilaminar blastocyst. With the electron microscope it can be seen that the cells are joined to each other by junctional complexes between the cell membranes. These effectively close off the space inside (Fig. 2.7c), containing the now liquified yolk body, from the outer environment, so that all subsequent movement of substances must be through the cells themselves.

This pattern of development differs in two important ways from the eutherian pattern: first, the eutherian egg does not cleave regularly to form two rings of cells but instead forms an uneven cluster or morula without a central cavity containing a yolk mass; second, junctional complexes develop only between the outermost cells, so that when the blastocyst forms by absorption of fluid some cells are trapped inside. This inner cell mass is destined to become the embryo proper while the outer cells form the invasive trophoblast which implants into the uterine wall. The marsupial blastocyst at the unilaminar state is not so differentiated, nor do these cells invade the uterine tissue, being prevented from so doing by the shell membrane.

The absence of a yolk body and the early differentiation of embryonic and extra-embryonic tissue in the Eutheria are considered to be advanced adaptations for intra-uterine life not developed by the marsupials; they still retain relics of a developmental sequence appropriate to a large yolky egg like the monotremes. However, the precise nature of the yolk body is not known, nor whether it serves an essential nutritional role in early development.

At the unilaminar stage all the cells resemble each other and no particular parts can be distinguished as the presumptive embryonic area, as can the inner cell mass in the eutherian blastocyst. In the next phase of development, however, the blastocyst develops a polarity that shapes its future. In one hemisphere some cells become detached from the protoderm and lie free in the blastocoele. In *Didelphis*[164] these endoderm mother cells enlarge greatly before proliferating, but in other species (Fig. 2.7d) they remain of similar size to the protoderm cells. In all species examined they spread around the inner surface of the protoderm but unlike that layer do not form a continuous sheet but rather a loose network held together by thin pseudopodia. This way of forming the endoderm is very similar to the way it is formed in the chick and is different from the eutherian manner. Also like the chick, the endoderm cells appear to induce the overlying protoderm cells to differentiate into an oval plate of thick cuboidal cells, which will become the medullary plate of the embryo. At the same time as these developments are taking place the liquefied yolk has been consumed and the outer mucopolysaccharide layer has diminished as it also is used up. As the blastocyst expands the protoderm cells become very thin but still retain complete attachment with each other. The cells of the medullary plate, however, do not expand so that this area becomes more distinct and is clearly visible in living vesicles. By the time that endoderm cells have spread right around the inner surface the mucopolysaccharide layer has disappeared and only the shell membrane persists.

The medullary plate now develops a bilateral symmetry with the appearance of the primitive groove and Hensen's node, in the same manner as does the chicken egg at 18 hours incubation. As in the chick, the middle germ layer or mesoderm proliferates laterally from the groove between the protoderm and the endoderm beneath. Ahead of the groove the notochord is formed and over it develops the neural tube from the cells of the medullary plate. In all of this the marsupial, exemplified by *Didelphis*,[164] follows the amniote pattern. From the neural tube develop the brain vesicles and spinal cord and from the lateral mesoderm differentiate the somites. These first appear on the eighth day in *Didelphis*,[164] on the 12th day in *Trichosurus*,[234] and on the 16th day in *Setonix*[234] and the tammer, a point of some interest to which we will return in a later section (p. 68).

Foetal membranes and placental transport

While the main form of the embryo is being laid down, mesoderm is spreading out beyond the limits of the medullary plate, until it extends nearly half way around the vesicle. However, it never reaches further, so that the marsupial yolk sac wall is in its lower half two layered and in its

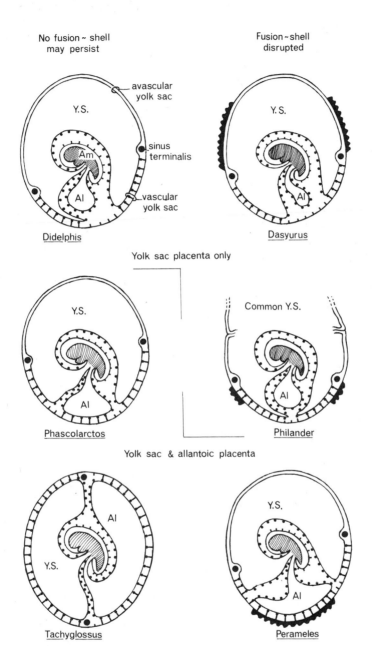

Fig. 2.8 Disposition of the foetal membranes in five marsupials and in the monotreme egg, after laying. The several patterns differ according to the proportion of the chorion that is vascularized, and hence respiratory; to whether or not the allantois participates in this; and according to the region, if any, of the chorion that invades the uterine epithelium. (From Enders and Enders,[86] Flynn,[89] Hill,[120] McCrady,[164] and Semon[230])

upper pole three layered (Fig. 2.8). As it is in the mesoderm only that blood vessels develop, this latter part becomes the vascular yolk sac, while the former is the avascular yolk sac.

Within the limits of the embryo and for a short distance beyond, the mesoderm splits into two layers, the somatic mesoderm being applied to the ectoderm and the splanchnic mesoderm to the endoderm. The space bounded by these layers is the coelom. Folds of the somatopleur now rise up at the head end and sides of the embryo and enshroud it within two membranes, an outer chorion and an inner amnion, which encloses a fluid filled space, the amniotic cavity. Once again, this manner of forming the amnion and chorion closely resembles the bird, reptile and monotreme manner and is distinctly different from the various means of formation by eutherian species.

In all amniotes another sac develops from the gut posterior to the yolk sac. This is the allantois, which in oviparous species enlarges greatly and comes to lie against the chorion and, by its vascular supply, acts as the main respiratory surface for the embryo. In Eutheria the allantois retains this function and also has a nutritive role as the foetal component of the definitive placenta. In most marsupials, however, it is an inconspicuous vesicle with a modest vascular supply buried in the enfolding yolk sac, and its only function is probably as a storage organ for excretory products. In a few species it is larger and in the peramelids it impinges upon the chorion and becomes highly vascular (Fig. 2.8). In the monotreme, *Tachyglossus*,[230] it becomes very extensive after the egg has been laid and is being incubated in the pouch.

Most of the endogenous resources of the embryo are used up by the time it has become a unilaminar blastocyst, and the last of the outer coat of mucopolysaccharide disappears during the phase of rapid expansion of the bilaminar vesicle. The shell membrane still persists and the vesicle lies free in the uterus, so clearly the material which passes into it and the material for synthesis of the new tissues must derive from the secretions of the uterus. It will be recalled that secretions pour from the glands during the luteal phase of the cycle, which coincides with the time of blastocyst expansion.

The yolk sac

In the tammar, as we have seen, the protein constituents of the uterine secretion differ from serum in several particulars, which is further evidence that it is not a simple exudate but is the product of active secretion. With the electron microscope the outer surface of the yolk sac wall of the tammar is seen to carry a thick weft of microvilli with pinocytotic vesicles at their bases. Mitochondria are numerous, as also is rough endoplasmic reticulum, all of which tends to support the concept of transport and metabolism of uterine secretions. A similar pattern is seen in the bilaminar yolk sac wall of *Philander opossum*,[86] and in both species, when the shell membrane disintegrates the microvillous surface of the embryo is presented directly to the uterine lumen and to the microvilli of the uterine epithelium with which a tenuous contact is made. In the tammar, interdigitation of the microvilli occurs and may be important in establishing a firm hold; earlier stage embryos can be rolled unharmed from the opened uterus, but this is not possible after the shell membrane has gone. In *Philander*, a still closer attachment occurs in the marginal zone between the vascular and avascular yolk sac walls (Fig. 2.8). Here the ectodermal cells are columnar and they form caps overlying the tips of endometrial folds. Giant cells develop and penetrate between the uterine cells into the stroma and they may form junctional complexes with the uterine cells. Since they do not penetrate capillaries or otherwise destroy maternal tissue, it seems likely that their function is to anchor the yolk sac closely to the uterus and thereby facilitate exchange between the two blood systems.

In *Dasyurus* a more profound attachment has been described in the marginal zone during the closing stages of gestation[120] (Fig. 2.8). With the light microscope an apparent syncytium of ectoderm cells unites with and invades the uterine epithelium and maternal capillaries, leading to extravasated blood entering the yolk sac. In the light of the work on *Philander*, this needs to be re-examined with the electron microscope to see if the cell membranes do really disappear or whether this also is primarily a bonding device.

Whether similar bonds occur in other marsupials remains to be discovered; it is unlikely that they occur in *Didelphis*, since the shell membrane remains intact and the vesicles can be removed undamaged at any stage of the short gestation, but a bonding similar to that in *Dasyurus* may occur in the koala, *Phascolarctos cinereus*.[12]

By the time of attachment the vascular part of the yolk sac is established and the blood is circulating. In contrast to the bilaminar part, it has a thinner ectoderm with fewer microvilli and the vitelline blood vessels come close to the surface as a result. In this region maternal capillaries are also close to the surface due to thinning of the epithelium, so that the two blood supplies are spatially close together, although the number of intervening cell membranes is unchanged. In the tammar, these histological changes are reflected in changes in the constituents of the yolk sac fluid.[211]

In the pre-attachment vesicle the protein concentration in this fluid is much lower than in serum, and the electrophoretic pattern is very similar to that for uterine secretion (Fig. 2.9); an albumin fraction is the main protein, with a small fraction which may be γ globulin, but the

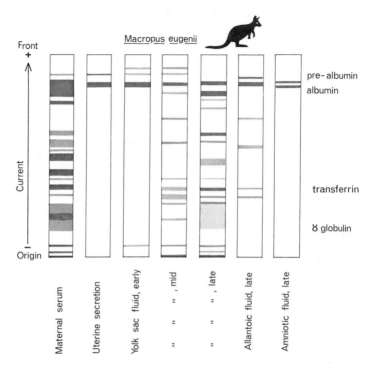

Fig. 2.9 Protein components of maternal and foetal fluids of the tammar, separated by disc electrophoresis in acrilamide gel. Note the increasing complexity of yolk sac fluid at successively later stages of gestation compared to the allantoic and amniotic fluids. (From Renfree,[211] with additional unpublished data by permission of the same author)

majority of serum proteins are conspicuously absent. By contrast, most serum proteins are present in the yolk sac fluid after attachment and it is tempting to suppose that they have been transported across the vitelline placenta; it is, however, equally possible that they have been synthesized by the embryo itself. It is important to know if maternal proteins are incorporated unchanged, since some of them have specific activities. The γ globulin fraction includes the antibody molecules or immuno-globulins which could confer passive immunity on the new born young if they are transferred unchanged but would not do so if they have been synthesized by the embryo. Transferrin proteins may convey ferrous ions from the mother to the embryo and these are essential for the synthesis of haemoglobin.

It was suggested by Hill[120] that the syncytial attachment in *Dasyurus* facilitated iron transfer, but in *Philander*[86] a reaction for iron was obtained in the endodermal component of the vascular yolk sac wall which could be due to synthesis or transfer.

In contrast to proteins, the tammar's yolk sac fluid contains free amino acids at 10 times the concentration found in maternal serum, and the constituent amino acids are qualitatively different in the two fluids.[211] Since amino acids are undetectable in uterine secretions, this could mean that the yolk sac wall is differentially permeable to the several amino acids or that hydrolysis of the albumin is taking place within the embryo prior to incorporation into new proteins. The presence in the yolk sac of some amino acids that are not found in serum supports the idea of de novo synthesis by the embryo.

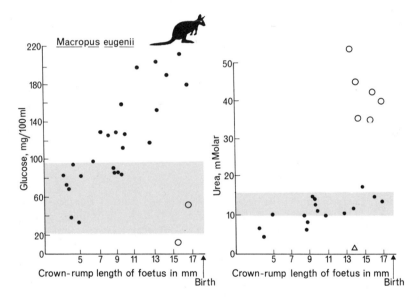

Fig. 2.10 Glucose (a) and urea (b) concentrations in yolk sac (●) and allantoic (o) fluids, at successive stages of pregnancy in the tammar, *Macropus eugenii*. Shaded areas represent the range of concentrations in maternal serum. It is evident that only in late pregnancy do the concentrations differ from maternal serum. (From Renfree,[211] with additional unpublished data by permission of the same author)

In the pre-attachment vesicle glucose concentration is no different from that in serum but it rises 3 to 4 fold after attachment (Fig. 2.10a), whereas fructose concentration does not change.[211] The tammar, like other macropods, has a pseudoruminant digestion characterised by low glucose and high volatile fatty acid levels in the blood (see Chapter 3). In the true eutherian ruminants, which have the same pattern, hexose sugar, in the form of fructose, is maintained at a higher level in the foetus than in the mother. It is therefore very interesting to see another parallel development here in the tammar embryo, which also maintains a hexose gradient. The main conclusion from these interesting results is that the wall of the tammar yolk sac is not an inert barrier but a functional membrane controlling the transport of a wide range of materials, some by diffusion, some by active transport.

The question that remains open is the relative importance for nutrition of the bilaminar, non-vascular, yolk sac and of the vascular yolk sac. If the bilaminar wall is the main route of absorption throughout gestation and the vascular yolk sac is essentially respiratory, it becomes clear why the entire surface does not become vascularized. Point is given to this argument by comparing the development of the foetal membranes in the egg of the monotreme, *Tachyglossus aculeatus*.[102] The endogenous yolk is only sufficient to sustain development to the bilaminar vesicle; thereafter, there is a rapid enlargement which, as in marsupials, is made possible by absorption of uterine secretions. During the intra-uterine phase, only part of the yolk sac is vascularized, as in marsupials. The embryo has about 19 somites when the egg is laid and development continues in the pouch without additional nourishment until hatching. During this phase the allantois is large and unites with the serosa so that the entire surface is vascularized (Fig. 2.8); its sole function must be respiratory.

The allantois

Both ammonia and urea occur in the yolk sac fluid of the tammar,[211] the former being at highest concentration before attachment and the latter afterwards. The urea concentration after attachment is the same as in maternal serum, which suggests that it diffuses out readily or at a rate commensurate with its production in the embryo. The urea concentration in the allantois, however, is about 3 times as high as in serum (Fig. 2.10b). This suggests that it acts as a site of urea storage in late gestation, although, since it is of small size, its function must be subsidiary to the yolk sac. In *Didelphis, Philander, Dasyurus, Pseudocheirus*[136] and *Phascolarctos*, however, the allantois becomes relatively much larger at the close of gestation and it would be interesting to know more about its function in these species. In *Didelphis*, the allantois reaches its maximum size 12–24 hours before birth and then shrinks coincident with the opening of the cloaca. McCrady[164] concluded that the progressive enlargement is due to accumulation of urine from the mesonephric kidneys and that when the cloaca opens, the fluid is discharged into the amniotic cavity, which in the closing stages of gestation appears rather bloated. However, in *Philander*, where it is very well vascularized, the Enders[86] tentatively suggest that fluid re-absorption may take place through these vessels, thus concentrating the stored urea.

The ultimate development of the allantois in marsupials is the fully developed chorio-allantoic placenta in species of *Perameles*.[89] In the early stages of gestation a vascular yolk sac develops, as in other marsupials, and becomes applied closely to the uterine epithelium, but subsequently the allantois makes contact with the chorion and vascularizes it (Fig. 2.8). In this region the ectoderm of the chorion proliferates a superficial layer of cells which fuse together to form a syncytium. Where the syncytium touches the uterine epithelium it erodes it. However, corresponding changes in the uterus have preceeded this stage so that the contact surface is in large part itself a syncytium, but a passive one. This change occurs in the non-pregnant uterus also but the subsequent changes are induced by the foetal syncytium. It erodes the maternal syncytium, exposing the necks of the glands and the subepithelial capillaries. The two syncytia fuse so intimately that in the first descriptions of the bandicoot placenta it was mistakenly thought to be a single layer. At the same time the deeper layer of cellular ectoderm in the foetal part is transformed into syncytium so that foetal capillaries are carried into the common syncytium and come to lie close to the exposed maternal capillaries, the blood being separated by the walls of the two capillaries alone. There is no breakdown of maternal capillaries as described in *Dasyurus*, but the transport of material each way must be very effective and might be supposed to confer great advantages on the bandicoots. The young at birth are more advanced than other marsupials at this stage and the nursing period is relatively short, while the gestation period of 12 days is shorter than in all other marsupials. Taken together it seems to suggest that these species have the most efficient placenta in the sub-class. Much debate has centred on whether this indicates that the bandicoots have retained an allantoic placenta from the common ancestor with Eutheria, or whether it has been independently derived from a pattern similar to *Philander*. The earlier embryologists could not believe that such a complex structure could have evolved independently twice in the mammals but, with our greater knowledge of marsupials and of viviparous reptiles today, this seems no less likely than many other physiological and anatomical convergences. If, then, it is to be considered an adaptation evolved in the Peramelidae, and evidently a very efficient one, it may be asked why the bandicoots, having developed an advanced allantoic placenta, do not also show the prolonged gestation with intra-uterine growth of the foetus, which has been universally adopted by Eutheria?

One view, long held, is that marsupials have not evolved a means of extending the active life of the corpora lutea, without which the uterus cannot be maintained in a luteal condition. But in fact, as we have seen, the corpora lutea in bandicoots are prolonged in an active phase during lactation. A more recent idea is that the crucial adaptation leading to the evolution of prolonged gestation in Eutheria is the factor, as yet undisclosed, by which the pregnant animal is prevented from mounting an allograft reaction against the foreign tissues of the embryo and rejecting it as rapidly as it would a foreign skin graft. Intra-uterine development is seen as the resultant of a series of compromises between the need to reduce the foeto-maternal barrier for more effective transport and, simultaneously, to retain it intact as a protection against allograft rejection. Several aspects of marsupial gestation suggest that this factor was not evolved in monotremes or marsupials: the persistence of the shell membrane for a relatively large part of gestation, during which time embryogenesis is very slow; a short period when close apposition of the blood streams occurs and the rest of embryogenesis, essential to a separate existence, is completed; and the general absence of intimate contact between foetal and maternal tissues. In this context the very intimate attachment of the bandicoot placenta may have conferred advantages in efficient transport, but may also have necessitated a shortening of gestation to avoid allograft reaction. Thus bandicoots have not been able to exploit one adaptation for lack of another, and the result is the birth of young unmistakably marsupial in their immaturity.

We return therefore to the point where this discussion began in noting that new born marsupials are remarkably uniform in size and development despite wide differences in adult size and total gestation length. We have seen that the disparities in gestation length relate in the main to differences in the length of the unattached phase when the shell membrane is still present, whereas the phase of embryogenesis, beginning with the appearance of somites and the vascular yolk sac, is of a relatively uniform duration and leads to a fairly uniform product. Because the embryo fails to invade the endometrium profoundly, as do the embryos of most Eutheria, the marsupial embryo is entirely dependent, until vascular contiguity is established, upon adequate secretion from uterine glands for its growth and development.

HOW IS SYNCHRONY BETWEEN MOTHER AND EMBRYO ACHIEVED?

Among mammals it seems to be a fact that intra-uterine development once started cannot be slowed down or stopped without imperilling the embryo. It is thus of great adaptive significance that the initiation of blastocyst growth be synchronized with uterine development appropriate to its present and future needs.

This is emphasized by the occurrence, in species of at least six different orders of Eutheria,[85] of a natural delay of development at the blastocyst stage, commonly called delayed implantation. The factors involved in delayed implantation are as varied as the species that display it, but all have in common the fact that the stage of embryonic development arrested is the unilaminar blastocyst which is the stage that the embryo can reach on endogenous reserves.

Since 1954 a similar embryonic stasis at the blastocyst stage has been found to be widespread in macropodid marsupials. It will be recalled that in these species parturition occurs in the pro-oestrous phase of the cycle (Fig. 2.5) and that suckling, while not inhibiting oestrus and ovulation, prevents development of the corpus luteum and the luteal phase in the uterus. If conception occurs at post-partum oestrus in any of these species the fertilized egg develops to the unilaminar blastocyst stage of 80 cells and then ceases to undergo further cell division or growth. This is analogous to diapause in insects and the circumstances under which it occurs and the factors involved in the resumption of development have been objects of much recent work. The marsupial system actually affords a more convenient means of studying the question of embryo-uterine synchrony than the eutherian, since the complexities of implantation do not supervene immediately after the end of diapause.

Natural occurrence of diapause in marsupials

Diapause was first discovered in the quokka, *Setonix brachyurus*.[240] This species breeds in the summer and birth is followed by post-partum oestrus and conception. During the early months of the nursing period most lactating females are found to be carrying a diapausing blastocyst 0·25 mm in diameter but as the season progresses this proportion declines, presumably through death and degeneration of the blastocyst itself. However, a few survive and at the time when the young one leaves the pouch in August another one may be born. By keeping lactating females away from males after February it can be demonstrated that the second offspring in August is indeed derived from a blastocyst that has been arrested in its development for 6 months. By artificially removing the pouch young from batches of females at monthly intervals it was shown that the proportion of these that produced a second offspring declined from 80% in February to less than 1% in August.[251] The quokka is a seasonal breeder and by August most females have already entered anoestrus, so that the arrested corpus luteum formed at post-partum oestrus, never completes development and has disappeared by the next breeding season.

In the tammar, *Macropus eugenii*, which is also a seasonal breeder, there is a similar decline in the proportion of females that will produce a second offspring after removing the pouch young, but in this species neither the arrested corpus luteum nor the blastocyst degenerate. By the time the first young vacates the pouch in September, seasonal anoestrus has supervened, but the corpus luteum and blastocyst continue in an arrested state until the next breeding season in January, when they both finally resume development. This has been dramatically demonstrated by segregating female tammars from males shortly after they had given birth in February and having them give birth eleven months later.[240] The only eutherian species known to have a diapause approaching this extent is the European badger which, like the tammar, conceives post-partum and can retain unimplanted blastocysts until the end of the same year.[85]

In young tammars lactation may not be necessary to initiate diapause. If a female reaches sexual maturity in June when the breeding season is closing it may be able to come into oestrus, ovulate and conceive, but the corpus luteum may fail to develop. Such animals also retain a diapausing blastocyst until the onset of the next breeding season.

In this species, then, the proximate factor for diapause can be either the suckling stimulus or some external cue, most likely photoperiod.

The red kangaroo, *Megaleia rufa*, which breeds all year round, also exhibits embryonic diapause. A high proportion of lactating females carry a diapausing blastocyst, but unlike in either of the preceding species most of these survive the 230-day pouch life of the previous offspring and are born a day or so after it leaves the pouch permanently. This synchrony indicates that development of the blastocyst must resume about 30 days before the pouch will be vacant. During this time the older offspring is spending much of its time out of the pouch eating grass and suckling less and less of the time. From these observations Sharman[236] concluded that the main factor inhibiting the blastocyst is the intensity of the suckling stimulus and he demonstrated this very elegantly in the following experiment, using female kangaroos that had just recently given birth to young, had mated post-partum, and were still suckling a large offspring out of the pouch. If the small pouch young was removed the diapausing blastocyst would resume development and be born 31–32 days later, despite the occasional suckling of the large offspring outside the pouch. If he first fostered a second young animal to such a female and then removed the small pouch young the next birth was delayed and occurred 34 to 52 days later. Thus the increased suckling stimulus provided by the two young outside the pouch delayed the onset of blastocyst development.

	(a) Detailed study of reproduction	(b) Young born after removing pouch young	(c) Blastocyst in uterus of lactating female
Burramyidae		Cercartetus concinnus	Acrobates pygmaeus
Macropodidae-Potoroinae	Bettongia lesueur Potorous tridactylus	Bettongia penicillata	Bettongia gaimardii
Macropodinae	Setonix brachyurus Megaleia rufa Macropus eugenii Macropus giganteus Wallabia bicolor	Lagostrophus fasciatus Macropus robustus Macropus rufogriseus	Thylogale thetis Thylogale billardierii Macropus irma

Fig. 2.11 Summary of the evidence for embryonic diapause in marsupials. (After Hill,[121] Sharman and Berger,[240] and Tyndale-Biscoe[278])

Embryonic diapause has not been so well studied in other species. However evidence from either the presence of blastocysts in lactating females, or the birth of young to females isolated from males after removal of previous pouch young, indicate that the phenomenon occurs in the Burramyidae and in four species of rat kangaroo, in rock wallabies, pademelons, hare wallabies and in all the larger wallabies and kangaroos (Fig. 2.11). In the grey kangaroo, *Macropus giganteus*, oestrus and ovulation are suppressed by the prolongation of gestation (Fig. 2.5) so that the opportunity for conception and blastocyst formation does not arise. However, towards the end of the lactation period, which is very long in this species, oestrus sometimes occurs. If conception ensues at this time, the embryo is arrested[64] at the blastocyst stage until emergence of the young one from the pouch, as in the red kangaroo.

Pituitary and ovarian factors involved in diapause

From these observations the immediate factor in diapause appears to be the arrested development of the corpus luteum and this in turn appears to result from a lack of pituitary stimulation. In the rabbit and rat it is known that there is an inverse relationship between the secretion (by the anterior pituitary) of gonadotrophins, and the secretion of prolactin, required for the maintenance of lactogenesis. The stimulus of suckling inhibits release from the hypothalamus of gonadotrophic releasing factor (GRF) while blocking the release of prolactin inhibitory factor (PIF). Oxytocin, released by the posterior pituitary has also been implicated in the stimulation of prolactin secretion and blockade of gonadotrophic secretion, and certain depressant drugs such as reserpine and chlorpromazine, by blocking the centre for GRF, inhibit ovarian function and stimulate lactation.

Based on these ideas the working hypothesis that John Hearn[116A] began with was that gonadotrophin secretion would rise after the pouch young was removed and that hypophysectomy at this time would prevent resumption of corpus luteum and embryo development. In the event, however, he found that while hypophysectomy caused lactation to stop in two days, the corpus luteum and embryo resumed development. Furthermore, he could not detect any rise in circulating gonadotrophin of intact females after removing the pouch young. Pursuing this further he found that if he removed the pituitary from female tammars during seasonal anoestrus when they had finished lactating, their quiescent corpora lutea and embryos would resume development immediately. Thus it seems that the corpus luteum of the tammar is not awaiting a gonadotrophic stimulus from the pituitary but is being tonically suppressed by it. A candidate for this role may be oxytocin, secreted by the posterior pituitary, and it is interesting that oxytocin injections to female red kangaroos, begun when the young was removed from the pouch, delayed the resumption of development for as many days as it was administered.[237]

Much more is now known about the next step in the chain of events, namely the interaction of the corpus luteum, uterus and blastocyst. Removal of the pouch young of the quokka early in the year leads to a fairly precise sequence of events.[274] For two days thereafter no detectable change occurs histologically but by day 4 the corpus luteum has begun to grow by division and enlargement of cells and between day 4 and day 6 a significant enlargement of the blastocyst has taken place. By day 6 the endometrium is developing the luteal condition. Similar changes have been observed in the tammar and red kangaroo, where also it has been observed that cell division resumes in the blastocyst and that the supernumerary sperm, which have been preserved in the mucin coat (Fig. 2.7) since fertilization, suddenly disappear. The most interesting feature of this sequence, however, is that histological changes in the blastocyst appear to precede those in the uterus and that both tissues follow the corpus luteum. If the corpus luteum is removed in the quokka on or before day 2, diapause is uninterrupted and the changes described above do not occur. If it is removed between day 3 and 6 the blastocyst resumes development but collapses as a vesicle, probably because the luteal phase fails to develop in the uterus. If the corpus luteum is removed after day 6 the luteal phase appears and embryonic development proceeds apparently normally to term. These results reinforce the idea that the blastocyst resumes development under the influence of the corpus luteum before the uterus does, but that unless uterine development follows, the blastocyst must perish because it cannot revert to the diapause state.

Progesterone injections given to lactating kangaroos or lactating or anoestrous tammars will induce blastocyst development[240] and, if continued for ten days, will develop the uterus sufficiently to maintain the embryo. Thus the role of the corpus luteum is clear: it must be stimulated to secrete progesterone or a similar steroid that is necessary for blastocyst and uterine development. The next question is whether the blastocyst resumes its development as a result of direct stimulation by progesterone upon its own cells, or merely responds to the alteration of the uterine milieu brought about by endometrial changes. If the former is true, then one of the functions of the corpus luteum would be to synchronize the development of the embryo with that of the uterus.

These questions have been examined in the quokka and the tammar by use of the technique of blastocyst transfer.[276, 280] By this technique a blastocyst is flushed from the uterus of a donor animal in sterile saline and is taken up into a micropipette. A non-pregnant recipient animal is anaesthetized, the uterus exposed, and the blastocyst inserted through the wall, still in the pipette, and flushed from there into the lumen. The recipient is sewn up and is examined two or three weeks later to find out how the embryo has developed. In the quokka it was found that diapausing blastocysts would successfully resume development in the uteri of recipients up to 6 days more advanced in development than the donor, but that blastocysts only 2 days more advanced than the recipient failed to develop. This supports the idea that the blastocyst cannot tolerate any delay in uterine development once it has resumed its own development, or that the quiescent uterus is inimical to a blastocyst after diapause.

To examine the question whether the blastocyst is directly stimulated by the corpus luteum a different design of experiment was used in the tammar. It was known that development would proceed normally after double ovariectomy on day 8. Transferring blastocysts into a day 8 recipient ovariectomized at transfer should therefore provide the blastocyst with a luteal uterus but no corpus luteum secretions. If it needs direct stimulation, a diapausing blastocyst should fail to develop in this recipient, whereas a day 8 blastocyst should be able to do so, having received its stimulus in the donor animal before transfer. In the event both kinds of blastocyst developed in both intact and ovariectomized recipients, thus demonstrating that the blastocyst does not require direct stimulation by the corpus luteum. However, it is possible that an inhibitory factor in the uterus of lactating animals is removed or inactivated by corpus luteum secretions, and that this enables the blastocyst to resume development. This would reconcile the results of this experiment with the observation that the blastocyst resumes development before the uterus. It also gets support from two other experiments. Dr. Meredith Smith found that tammar blastocysts transferred from the uterus to a porous (millipore) capsule and returned to the body cavity of the still-lactating mother resumed development, suggesting that they had escaped from an uterine inhibition. She[260] also found that injections of oestrogen to lactating tammars would induce resumption of blastocyst development but not a luteal type of endometrium.

These several studies demonstrate that embryonic survival in these species beyond the blastocyst stage is completely dependent upon the onset of the luteal phase in the uterus, and that the corpus luteum provides a secretion which synchronizes the two phenomena.

Several theories have been propounded to account for diapause in macropods; some are ecological and will be considered in Chapter 3, others suggest that it is a fortuitous concomitant of extending the gestation length to fill the length of an oestrous cycle and that a longer gestation is the main selective advantage.[238] We have seen already (Fig.

2.5) that there is no substantial difference in size or development between the young born after a short gestation and those born after a longer one, as might be expected if the latter were of selective advantage. Furthermore, this view does not take account of the fact that the blastocyst itself must be specifically adapted to a long survival in diapause, an adaptation encountered in a wide range of genera in both sub-families of the Macropodidae. Such a widespread adaptation must have a greater significance in marsupial reproduction; and the significance is not hard to see. It is clearly of paramount importance that the embryo develop in a luteal uterus for it will perish if it does not do so. Even a few days asynchrony is fatal once development has resumed, so an adaptation that enables a stasis of development and ensures embryo-uterine synchrony would have great selective value and might therefore be expected to occur in all marsupials. If this be so, the diapause found in macropodids is not unique to this group of marsupials but is a normal stage of development in all marsupials; the unusual aspect of the macropod phenomenon is that the suckling stimulus inhibits the signal which should end diapause so that this is greatly extended.

What evidence is there that a similar diapause occurs in other marsupials? The observation that young female tammars, which conceive at the end of the breeding season, enter anoestrous and that their blastocysts diapause[240] is suggestive evidence that the phenomenon does occur normally in the absence of lactation.

It was mentioned earlier that gestation lengths are variable even between related species. One particularly marked comparison can be made between the rat kangaroos, *Bettongia lesueur*[278] and *Potorous tridactylus*. The adults of both species weigh about 1 kg and both species give birth to single offspring that weigh 300 mg, and are suckled for 120 days. Yet the gestation period of *Bettongia* is 21 days while that of *Potorous* is 38 days (Fig. 2.5). How is it that *Potorous* needs 17 more days to produce the same sized offspring as *Bettongia* unless embryonic development is deferred for a longer time in the former species? At present the intra-uterine development of neither species is known but such information could provide crucial evidence for this hypothesis. Then again, the gestation period of *Antechinus stuartii* is variable when timed from oestrus but when backdated from parturition it is found that the growth of corpora lutea and embryogenesis are well synchronized and the variation is all in the early part of gestation when the embryo is a blastocyst. Among the Burramyidae the females of the pigmy possum, *Cercartetus concinnus*,[63] and the pigmy glider, *Acrobates pygmaeus*,[121] carry uterine blastocysts during suckling of a previous litter, and in the former species the second litter is born at the end of the first lactation. In *Cercartetus* some cell division does occur in the cells of the blastocysts but development does not progress beyond the blastocyst stage during lactation. Thus current knowledge is not at variance with the hypothesis that diapause may be a normal phase of marsupial development, evolved because it ensured synchrony of embryo and uterus; but this hypothesis still awaits test. If it should prove to be valid it would not necessarily contradict the evidence for diapause being an ecological adaptation. As with many characters, those selected for a primary advantage may confer secondary advantages to the possessor, which reinforce the initial advantage.

CONCLUSIONS

The main conclusions to be drawn from this first part of the chapter are that marsupial reproduction resembles that of the Eutheria in some particulars and the Monotremata in others.

Like most eutherian species, most marsupials are polyoestrous and the sequence of events in the oestrous cycle is similar, and is controlled by the same ovarian hormones. Also, the marsupial egg, although larger, is almost devoid of yolk and is therefore utterly dependent on uterine secretions for its nourishment and intra-uterine development.

The foeto-maternal relationships, however, have a closer resemblance to the monotreme pattern, and they are summarized in Fig. 2.12. In this comparison three phases are proposed: the autonomous phase, the absorptive phase, and the phase of differentiation.

The absorptive phase is of paramount importance and is dependent on the development of a secretory endometrium in the uterus. The preceding, autonomous phase is to be viewed as a preliminary phase, when the egg lives on its endogenous reserves. The echidna, with its relatively large yolky egg, is autonomous for longer than the marsupial and this phase merges into the absorptive phase. The marsupial has no such buffer but if the diapause observed in macropods is a common phenomenon, it could fulfil this function until uterine secretions are provided. The bilaminar yolk sac is probably the main route of absorption in the second phase, which is characterized by a rapid increase in size due to a filling of the yolk sac, by diffusion and active transport. This is not impeded by the surrounding shell membrane.

As differentiation of the embryo begins, the respiratory rate must greatly increase, and this is served by the vascular part of the yolk sac, which is seen to increase in area as gestation advances. If the vascular part is not also absorptive, its gradual spread must limit the further absorption of nutriment by the non-vascular part. Dissolution of the shell membrane at this stage may increase the transport capacities of both parts of the yolk sac, and hence allow longer intra-uterine development.

But the breakdown of the shell membrane may expose the foetal tissues to an allograft response by the mother and their rapid rejection. The echidna egg, however, does not lose its shell and is laid at the end of the

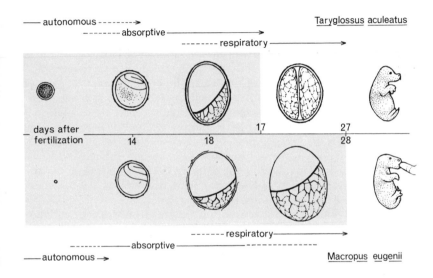

Fig. 2.12 Comparison of three phases in the development of the monotreme, *Tachyglossus aculeatus*, and the marsupial, *Macropus eugenii*. Intra-uterine stage shaded, pouch stage unshaded.

absorptive phase; its increasing respiratory needs during differentiation being met by a complete envelope of vascular yolk sac and vascular allantois under the shell.

In both the Marsupialia and the Monotremata the end result is a very small offspring, in which the organs are determined but only a few are functional. It is maintained in a pouch or abdominal depression, where the temperature and humidity are stable, and is nourished exclusively on milk. Pouch life is a period of rapid growth and development of the many adaptations for homeostasis. The second part of this Chapter is therefore concerned first with the adaptations of the neonatal marsupial and the supportive role of lactation, and second with the development of homeostasis.

HOW IS THE YOUNG SUCCOURED?

Parturition

Endocrine factors

The genital tract of marsupials is double as far as the urogenital sinus, which lies posterior to the bladder, but the two vaginae become secondarily united in front of the bladder at the level of the uterine cervices (Fig. 1.2). At birth the foetus leaves the uterus and passes into this common chamber or vaginal cul-de-sac and, forsaking both lateral vaginae, continues by a direct route through the connective tissue strand that connects the posterior wall of this chamber to the anterior wall of the urogenital sinus. In most Macropodidae and in *Tarsipes spencerae*, this pseudovaginal canal remains patent after the first parturition and becomes lined with epithelium derived from the cul-de-sac, but in all other marsupials the rough tear in the connective tissue is rapidly repaired, so that no evidence of it remains a day or so after birth. Thus two processes at least are involved in the delivery of the young; first, evacuation of the foetus and its membranes from the uterus, and second, opening of, and transport along, the birth canal.

There is no information at present on the way in which parturition is initiated in marsupials; certainly the ovaries are involved in some way since the process is prevented by ovariectomy performed late in pregnancy of the opossum,[115] brush possum and quokka.[275] It has been suggested that involution of the corpus luteum is the signal in species such as the brush possum and the opossum, but this cannot be the signal in the bandicoot, which gives birth before involution, nor in the macropods, which give birth long after. It seems likely, therefore, that at least two distinct types of endocrine control occur, perhaps related to the two conditions of the birth canal. Most of the species that have a transient birth canal give birth while the corpus luteum is still active and in two of these species, *Trichosurus* and *Didelphis*, progesterone induces a loosening and oedema of the connective tissue strand, similar to that observed at mid-cycle, when parturition occurs.[277] Conversely, the quokka, which has a patent birth canal and a corpus luteum far gone in its regression at the time of parturition, is nevertheless dependent on the ovaries for successful delivery. Oestrogen injections did not help ovariectomized quokkas to give birth[275] and the full term foetuses were either retained dead in the uteri or dead in the vaginal canals, whereas a relaxin preparation injected into similarly treated quokkas resulted in live births in some and live uterine foetuses in the others. It will be recalled that a relaxin-like activity has been assayed in possum corpora lutea but not in possum ovaries per se. If, however, relaxin is synthesized by quokkas and plays a role at parturition, it is unlikely that the corpus luteum is the site, nor can it be the new Graafian follicle, because some quokkas fail to ovulate at the end of a fertile cycle and yet give birth and suckle their young satisfactorily. The other ovarian tissue must therefore be considered the most likely site in this species, but so far it has not been assayed for relaxin.

Progesterone may be involved in preparing the birth passage in the connective tissue strand, while relaxin has a more subtle role in evacuating the uterus and facilitating transport along the birth canal. One might speculate that the extension of gestation in macropods into the post-luteal phase of the cycle was in part made possible by the evolution of a permanent birth canal and the reduced dependence on corpus luteum secretions for birth which that allowed.

The process of birth

Birth has now been observed and described for many species of marsupial and the story of its discovery often told.[245] In all species the young makes its way from the urogenital opening to the pouch by its own efforts in a matter of minutes. Before considering the adaptations that enable it to do so, the behaviour of the mother during the process is interesting.[241] The phenomenon of diapause, which can be terminated by removal of the pouch young, enables the time of birth in kangaroos to be programmed to the observer's convenience and a detailed series of observations of the entire behaviour leading up to birth in the red kangaroo, *Megaleia rufa*, have been made and recorded on ciné film.[58]

The pouches of non-lactating kangaroos contain a brown, dry scale and in the last week of pregnancy this is removed and the pouch appears clean and moist. It is accomplished by the female putting her muzzle into the pouch while holding the sides open with her fore paws. Non-pregnant kangaroos at the same stage of the cycle also clean the pouch in the same characteristic manner and also occasionally adopt the 'birth position'. In this the female sits on the butt of the tail with the tail and hind legs extended forward so that the urogenital opening is directed upwards. In the hour or so preceeding birth, pregnant animals adopt this posture much more frequently than non-pregnant animals and the intensity of pouch cleaning increases greatly. At the same time the animal begins to lick the urogenital opening frequently.

Birth is heralded by a flow of fluid, probably from the ruptured yolk sac, which is immediately licked up and is followed by the appearance of the intact allantois, which may fall to the ground. The young emerges head first, still enclosed in the fluid-filled amnion from which it now frees itself with its swinging, clawed arms. It is followed by the remaining yolk sac membranes, trailing away from the umbilicus, from which it breaks free, partly aided by the mother's persistent licking. If the mother is undisturbed, the young orientates itself towards the pouch and rapidly moves there by grasping the fur in its claws and by alternate movements of the forelimbs. At the same time the head turns from side to side so that the muzzle describes an arc. Within a few minutes, usually less than five, it has gained the lower lip of the pouch and thence passes from view. If the mother has not been suckling a previous offspring, all four teats are available, otherwise three are. These have developed a small apical bud which the young now sucks into its mouth. For two days after birth the young can be removed from one teat and placed on another and it will suckle successfully, but as time passes this fails because only the suckled gland secretes milk. Similarly, young transferred to the budded teat of a non-pregnant female 33 days post-oestrus, will attach and be suckled successfully to weaning. This technique can also be applied to inter-species transfer[178] provided the recipient female is at the stage of the cycle when parturition would normally occur in that species.

From the less complete observations on other marsupials the process of birth appears to be much the same as for the red kangaroo. The grey kangaroo gives birth in a standing position but the quokka, possum, koala and opossum adopt the same posture as the red kangaroo. In the opossum, however, none of the several observers of its birth have said that the young emerges still in the amnion, and indeed McCrady[164] expressly states that both amnion and allantois are ruptured before birth and the yolk sac wall is left behind in the uterus. Clearly, if the young kangaroo is born in the amnion it cannot participate actively in traversing the birth canal, whereas the young opossum, if free from membranes, could tear the loose connective tissue separating it from the urogenital sinus. Observations on other species representative of both types of birth canal would be needed to decide whether there is any real significance in this apparent difference between the opossum and the red kangaroo at birth.

Neonatal adaptations

The marsupial at birth is a marvellous composite of embryonic structures and precociously developed functional organs, the latter enabling it to reach the pouch, to respire and to gain nourishment from the mammary gland (Fig. 2.13). The fore limbs and shoulder region are well developed and the digits are armed with sharp recurved claws. The scapula is well formed in cartilage and is continuous with the coracoid and the cartilage of the sternum,[52] but a week after birth the coracoid articulation with the sternum is severed. In the opossum[62] and brush possum[286] the muscles of the shoulder and fore limb have the adult

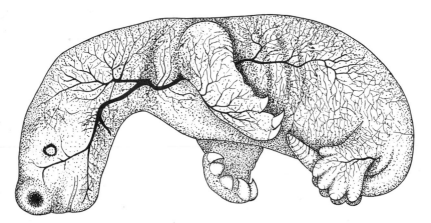

Fig. 2.13 New born unattached young of *Trichosurus vulpecula* showing the well developed fore limbs with claws and the prominent cutaneous blood vessels seen through the ntegument. (From Lyne[163])

disposition and the relatively large coracoid does not give origin to large coracobrachial muscles as its homologue does in the monotremes and reptiles; rather its functional significance seems to be to give firm support to the glenoid on which the fore limb articulates during the journey to the pouch. The muscles of the neck and thorax are also well developed and the young animal carries out a series of alternate lateral contractions, starting at the head and ending with the fore limbs. These are reflex movements, as Langworthy[151] demonstrated in the opossum (see p. 207), and are unaffected by excising the cerebral hemispheres of the brain. Indeed, cerebral response cannot be elicited until three weeks after birth and the pyramidal tract, which carries fibres direct from the cortex to the spinal cord, does not develop until the fifth week. Olfaction is probably the only sensory perception of the new born marsupial, since sensory cells are present in the nasal epithelium and the olfactory nerve connects to the olfactory lobe of the brain.[164] The eyes are embryonic, and balance and hearing do not develop for 3 and 8 weeks respectively. The lungs have simple alveoli and a thin epithelium covers the capillaries, so that gaseous exchange is probably not very efficient by this route. However, the new born possum and kangaroo have very conspicuous subcutaneous blood vessels (Fig. 2.13) and a moist skin, so that it is possible that respiration occurs across the skin as well as the lungs in these species.

   The buccal cavity is large and the tongue protrudes from the circular mouth. This arrangement results in the end of the teat expanding inside the mouth so that the young becomes firmly attached to it and can be removed only with difficulty. At the back of the mouth the epiglottis is large and extends through the soft palate so that the glottis opens into the nasopharynx. The buccal cavity extends around each side and communicates with the oesophagus. By this arrangement the young is enabled to respire concurrently with suckling. The stomach and duodenum are in an advanced stage of development at birth and so is the pancreas; in contrast, the small intestine and colon are not. The mesonephros is the functional excretory organ before birth, discharging into the allantois, and it continues to function for 2 weeks after birth, while the Wolffian ducts function as the urinary ducts.

Composition of milk

The main constituents of marsupial milk, as of eutherian and monotreme, are water, sugars, fats and protein (Fig. 2.14), but because of the long period of lactation the relative composition of marsupial milks changes more obviously in association with the developing young. The milks of the brush possum[108] and the red kangaroo[155] have been adequately sampled through lactation; less extensive observations on the opossum,[36] the quokka and the euro, *Macropus robustus*, have given similar results. The total solids are higher than in cow or human milk and the relative composition is different from them. There is a uniform amount of sugars but a progressive increase in the fat and protein content as lactation proceeds.

The reducing sugars make up a smaller proportion and, unlike in eutherian milk, lactose is not the main sugar of marsupial milk and represents less than 3%, while other oligo-saccharides represent about 5%, except in *Didelphis*. On hydrolysis this latter material yields galactose but no glucose. A galactan that yields only galactose would require only one enzyme instead of the two needed to hydrolyse lactose to glucose and galactose, which may be the significance of this in the nutrition of the small pouch young. However, in the intestinal mucosa of the only two pouch young so far examined, lactase activity was higher than in adult marsupials of the same species and similar to eutherian levels[143] (Fig. 4.2).

The lipid content of red kangaroo milk increases from 2% during the first 70 days of lactation to 10% later. The most abundant component in the lipid fraction is triglyceride[106] but traces of phospholipids, cholesterol and free fatty acids are found. Unlike eutherian milk, the triglyceride in early milk is largely composed of saturated palmitic acid, whereas later in lactation the unsaturated oleic acid characteristic of eutherian milk predominates. In Fig. 2.15 these differences are brought

Species	Water	Total Solids	Protein	Lipid	Reducing sugar	Other sugar
Human[289]	87·4	12·5	1–1·5	3–4	7–7·5	
Cow[289]	87	13	3–4	3·5–5	4·5–5	
Didelphis marsupialis[36]	76·8	23·2	8·4	11·3	1·02	0·6
Trichosurus vulpecula[108]	75·5	24·5	9·2	6·1	3·2	6·0
Megaleia rufa[155]	78	22	6–8	2–10	2	2–15

Fig. 2.14   Percentage composition of three marsupial milks compared to two eutherian milks.

out, as well as the fact that a red kangaroo can produce both kinds of milk simultaneously from adjacent glands. This comes about when she is suckling a small young in the pouch, as well as a large offspring outside the pouch (see later).

Milk protein contents of several macropods have been examined by electrophoresis, which has shown more clearly the components that contribute to the overall increase, determined by earlier work. In the quokka and red kangaroo, and in the opossum, the patterns for casein show a peak corresponding in mobility to the high phosphorus or alpha casein of cow's milk but lack the other caseins of cow's milk.[258] The whey

Kangaroo number	Age of young in days	Crude lipid, g/100 g milk	Palmitic acid content of triglyceride, g/100 g	Oleic acid content of triglyceride, g/100 g
1	40	1·6	47·9	22·5
	280	5·2	25·0	50·6
2	62	4·2	37·5	32·0
	300	8·3	23·6	49·1
3	120	5·4	29·1	42·3
	360	9·3	24·5	53·4

Fig. 2.15 Differences in the lipid content of the milks of red kangaroos, *Megaleia rufa*, being secreted simultaneously from adjacent glands to small pouch young and large young at heel. (From Griffiths, McIntosh and Leckie[106])

proteins are more complex and more interesting: the albumin and the beta and gamma globulin fractions of quokka milk[139] remain constant throughout lactation at levels below those in the mother's serum (Fig. 2.16), and it is thought that they are derived from the blood and not synthesized in the mammary gland. Since the concentration of albumin, which is the smallest of these molecules (M.W. 69 000), is only 15% of the serum level, whereas gamma globulin (M.W. 165 000) is 40% of the serum level, it is unlikely that they enter the milk by simple filtration but rather that there is selective transfer of the gamma globulin, as in Eutheria.

By contrast with these fractions, the alpha globulins are at higher concentration than in serum at all stages of lactation and rise to high concentrations in the latter phases of lactation. The main components

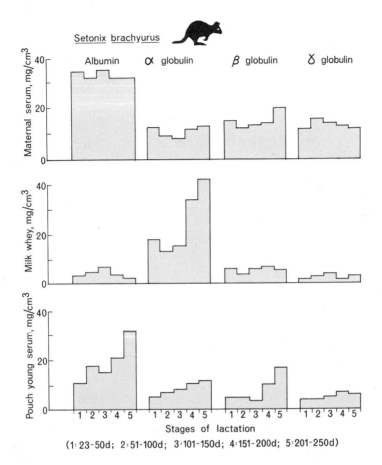

Fig. 2.16 Protein changes in maternal serum, milk whey and pouch young serum throughout lactation in the quokka. Proteins were separated by electrophoresis on cellulose acetate strips at pH 8.6. The large component of α globulins in milk whey after 151 days is probably the same moiety that runs ahead of albumin in starch gel electrophoresis and which is referred to as pre-albumin 1 and 2 in Fig. 2.17. (After Jordan and Morgan[139])

of this late rise have electrophoretic mobilities different from any serum component but resemble the specific proteins, β lacto-globulins and α lact-albumins, of cow's milk. In the quokka one of these proteins first appears in the milk at 150 days and the other at 200 days, which overlaps the period when the young quokka is respectively becoming homeostatic and is leaving the pouch and beginning to extend its diet to plant

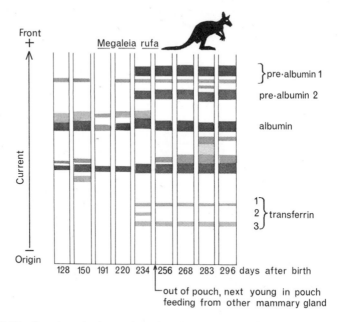

Fig. 2.17 Starch gel electrophoresis of whey proteins from milk of a red kangaroo during the second half of lactation, to show the change in composition just before the suckling leaves the pouch and the next young enters it. Thereafter the female produces early and late milk simultaneously from adjacent mammary glands. (From Bailey and Lemon[16])

material (see p. 144). Similar proteins appear in the whey of red kangaroos[16] at 220 days (Fig. 2.17) and in grey kangaroos[156] at 240 days and 260 days and, like the quokka, this coincides with the time of emergence from the pouch of each species.

In the red kangaroo, it will be recalled that the blastocyst carried by most lactating females comes out of diapause at this time and is born about 230 days after the previous young. The new born young attaches to one of the three unused teats in the pouch, while the other young suckles the much-elongated teat from outside the pouch. It thus comes about that the one female simultaneously secretes from adjacent mammary glands two kinds of milk, differing in lipid and protein content and composition that represent respectively the beginning and the end of lactation (Fig. 2.17). The endocrine control of this still remains to be explored.

All the protein components are at a low concentration in the serum of quokka pouch young[139] which are less than 50 days old (Fig. 2.16) and correspond to the low levels found in the blood of eutherian foetuses of comparable developmental age. Albumin rises progressively to adult levels by 250 days but the large component of alpha globulin in milk of late lactation does not appear at all in the young animals's serum and has presumably been digested. The gamma globulin concentration is low in the young but the beta globulin, which includes the iron-binding proteins or transferrins, rises progressively through lactation to a level well above that of the milk whey, which again resembles the pattern in peri-natal Eutheria.

One major difference in macropod marsupials and Eutheria, connected with this point, is the means of transport of iron and copper to the young. In Eutheria, such as man and ox, copper and iron are actively transferred across the placenta so that the new born has levels in the blood (in the transferrins) and in the liver much higher than those in the mother, but the milk transports very little of these elements (Fig. 2.18), so that the levels in the nursing young fall progressively until weaning. Now, although there are transferrin proteins in the yolk sac fluid of the tammar (Fig. 2.9), it is impossible that the marsupial at birth could contain adequate stores of copper or iron to maintain it through the long nursing period; indeed, in the young quokka the levels of both elements rise progressively during pouch life, as do the transferrin proteins in the plasma. These elements are obtained from the milk: during the first 170 days the concentration of iron in the milk is about five times[140] and of copper, three times,[22] the concentration of maternal plasma and then falls to maternal plasma levels when the young begins to emerge from the pouch (Fig. 2.18). The transfer of iron to the milk in the quokka is

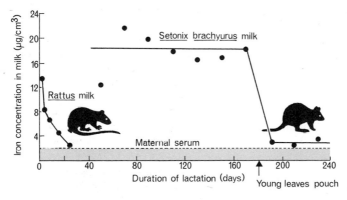

Fig. 2.18 Concentration of iron in the milk of the quokka and of the rat, compared to the concentration in maternal serum (shaded). (After Kaldor and Ezekiel[140])

not due to a greater permeability of the alveolar epithelium than in eutherian mammary glands, since raising the maternal plasma iron artificially did not affect the milk iron concentration;[141] nor is it due to a greater affinity of whey transferrins for iron than plasma transferrins have.[88] Quokkas, like red kangaroos, do occasionally suckle two young differing in age by about 170 days, so that, as with the secretion of different proteins and lipids from adjacent mammary glands, such an animal must be transporting a far higher concentration of iron across the newly suckled gland than across the other. Whatever the transport mechanisms are they must be intrinsic to the gland and be a function of the stage of development of the gland.

Lactation and the control of galactopoiesis

The stimulus of suckling itself and the amount of milk withdrawn from the gland probably constitute the main factors in the maintenance of lactogenesis. Thus in the non-suckled gland of the brush possum[235] shortly after parturition the alveoli were larger and had a thinner epithelium than the suckled gland of the same animal, which suggests that milk was being retained there under pressure and was inhibiting further secretion by the alveolar epithelium, as it is thought to do in Eutheria. The stimulus of suckling does however have a general effect, since the non-suckled glands of suckling possums regress more slowly than do those of non-lactating animals and milk can be expressed from them for a longer time.

Further evidence for the importance of the suckling stimulus rather than milk withdrawal in lactogenesis has been shown in the opossum. Lactation normally ceases at about 100 days when the young begin to feed actively outside the pouch. By fostering a second litter 60 days old to a weaning mother, Reynolds[213] was able to maintain lactation in this animal for 154 days. The mammary glands first regressed to the size appropriate to the younger litter and then began to enlarge again slowly in response to their increasing demands. The persistent suckling of the younger litter, although withdrawing less milk, was yet more effective in maintaining lactation than the infrequent suckling of the older litter. In two other examples the activities of the young determined the length of lactation. A pouch young of the swamp wallaby, *Wallabia bicolor*, was fostered into the pouch of a red kangaroo[178] where it grew at a faster rate than normal swamp wallabies. It emerged from the pouch at 267 days, which is the normal pouch life for a swamp wallaby but 30 days longer than that of a red kangaroo. Similarly, a grey kangaroo young, fostered into a red kangaroo, was retained in the pouch for 374 days, or 135 days longer than a red kangaroo young would have stayed. It would be interesting to know how the milk constituents are altered by these various manoeuvres. The accelerated growth of the swamp wallaby is particularly interesting because its rate of growth equalled that of young red kangaroos and greatly exceeded that of normal swamp wallabies, and it reached sexual maturity much before them. Did it draw more milk from its foster mother than it would have done from its own natural mother? Or is the milk of red kangaroos more nutritious and is it this that enables them to leave the pouch earlier than either the swamp wallaby or the grey kangaroo?

WHEN DOES THE YOUNG ACHIEVE HOMEOSTASIS?

Thermoregulation

During the first half of pouch life the young marsupial is in an environment with a high stable humidity, a stable temperature and, in species with a closed pouch, a CO_2 concentration of about 5%.

The body temperature of the young opossum removed from the pouch during this period rapidly changes to the ambient temperature;[213] if this is a low temperature the animal becomes torpid and may cease to move, but will as quickly recover if returned to the pouch. As it grows older the fall in body temperature is less precipitate and in the second half of pouch life it may hold a steady temperature for several hours, after removal from the pouch, before falling to ambient (Fig. 2.19). Finally, at about the time the young is emerging from the pouch it can maintain a stable body temperature against an ambient gradient as well as the adult animal can.

Maintaining the temperature of the body above ambient requires the animal to have an endogenous source of heat, and insulation sufficient to prevent its rapid dissipation. The net expenditure of energy can be assessed by measuring the oxygen consumed by the animal in a closed system, maintained at a given temperature. The pouch young of opossums and quokkas[249] have been subjected to this procedure and the results corroborate the previous results.

The oxygen consumption of quokkas less than 80 days old, and of opossums less than 60 days old, is directly related to the ambient temperature up to 35°C, whereas in animals older than this, which had been shown to be capable of thermoregulation, oxygen consumption was inversely proportional to ambient temperature (Fig. 2.19). In the opossum, oxygen consumption at low temperatures was greater in 82-day old than in 92-day old young, which may be due to the better insulation of the older animals that had developed a full pelage of under-fur and guard hairs. Similarly, in the quokka the pelage is complete at 165 days, which coincides with the development of thermoregulation. In this species the ability to shiver was found to develop first at 120 days,[247] which is the

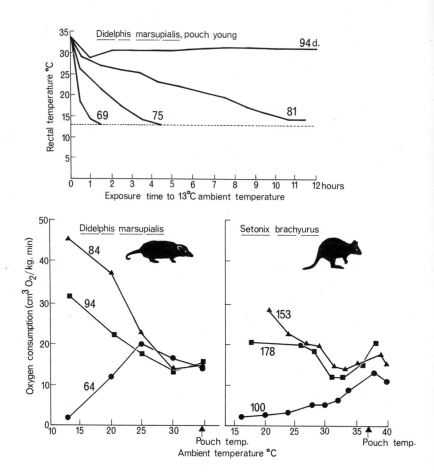

Fig. 2.19 Top. Responses to cold of pouch young opossums of increasing age. Thermal stability has begun at 81 days and is fully established at 94 days.
Bottom. Oxygen consumption of young opossums and quokkas shows the same pattern: before thermoregulation has developed, oxygen consumption is directly related to ambient temperature. Older animals become active at moderate temperatures, 25°–30°C, and then consume more oxygen. As thermoregulation is established the oxygen consumption is inversely related to ambient temperature; the highest values being observed before the young are fully insulated by fur. (After Reynolds[213] and Shield[249])

age when thermoregulation is beginning. The blood haemoglobin concentration, upon which the animal depends for oxygen transport, rises from 9 g/100 cm³ at 30 days to 13 g/100 cm³ at 180 days when thermoregulation is complete.

The thyroid gland is closely involved in the maintenance of homeothermy, and thyroidectomized mammals are unable to respond to cold temperatures by increasing their metabolic rate. The thyroid synthesizes two hormones, thyroxine and tri-iodothyronine, which both contain iodine as a component of the molecule. For this, iodine is sequestered by the thyroid and the uptake of radioactive I^{131} can thus be used to measure thyroid activity. When secreted into the circulation the general effect of these hormones is to stimulate most enzymic systems, such as glucose oxidation and protein synthesis; the uncoupling of oxidative phosphorylation may channel oxidative energy into heat rather than into synthesis and be the means by which the hormones stimulate endogenous heat production. This being so, the development of thermoregulation during pouch life might be expected to be associated with onset of thyroid function. This has been examined in the pouch young of

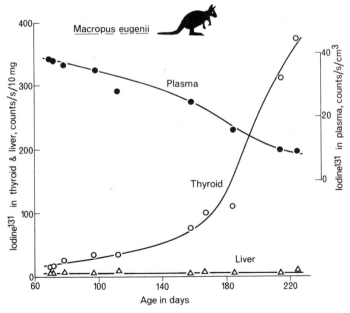

Fig. 2.20 Uptake of Na I^{131} by the thyroid, liver and blood plasma of tammar pouch young, injected 24 hours before at 1 μCi I^{131}/100 g body weight. (After Setchell[232])

the tammar, *Macropus eugenii*.[232] The same pattern of oxygen consumption was observed as in the quokka and opossum, but the change to homeothermy occurred at about 140 days.

The thyroid tissue differentiated into the characteristic follicles by 75 days but the typical accummulation of secretion in the follicles did not appear until 180 days.

The differentiation of the thyroid tissue and the onset of active synthesis, as determined by a sharp rise in its uptake of I^{131} (Fig. 2.20), coincided with this period. At the same time the level of I^{131} loosely bound to serum proteins fell, again indicating that the thyroid was sequestering all of the injected iodide. Pouch young thyroidectomized before 140 days, subsequently failed to grow at the normal rate, to differentiate and grow a full pelage, and were unable to respond to low temperature by increased oxygen consumption.

None of this work has considered the response of pouch young to heat stress when the ambient temperature exceeds the body temperature but, by the end of pouch life in both the opossum and the quokka, the preferred body temperature was about one degree lower than the temperature of the pouch (Fig. 2.19) and this may be the stimulus for the young to leave the pouch. Point is given to this idea by the observation in the opossum that young of small litters leave the pouch at the same age as young of large litters[213] and not later as might be expected if lack of space in the pouch were the primary stimulus to leave.

Kidney function

The high humidity of the pouch may be important, as mentioned earlier, for very small pouch young, to facilitate cutaneous respiration but it is probably not significant in the water balance of pouch young, as the skin is very impermeable, dehydration is very slow and continuous suckling of milk seems to more than balance water loss. Furthermore, young opossums can survive and recover after the loss of 25% of the initial body weight.[213] But both water and ionic regulation must be effective before the end of pouch life, when the animal is homeostatic. At birth the mesonephric kidney is the functional excretory organ but thereafter its function is gradually taken over by the metanephric kidney in which new nephrons are being differentiated in the nephrogenic zone at the periphery of the organ. In the quokka, *Setonix brachyurus*,[35] the rate of kidney growth is isogonic with the overall growth of the body until 100 days (Fig. 2.21), when the nephrogenic zone disappears and the kidney weight : body weight ratio starts to decline to the adult value of about 0·6%. However, while this is happening the physiological function of the kidney is developing and two anatomical features are important

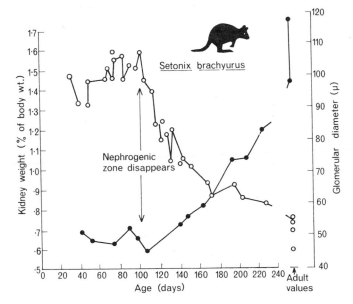

Fig. 2.21 Differentiation of the kidney of the quokka; as the animal grows the relative weight of the kidney declines as the nephrogenic zone disappears, but the size of the gomeruli increases as they become functional. (From Bentley and Shield[35])

in this: the diameter of the glomeruli at the proximal end of the nephrons begin to increase slowly (Fig. 2.21), which means that filtration across the glomerular membrane from the blood increases, and second, the loops of Henle increase in length; since it is in the loops that effective concentration of ions occurs, the young quokka gradually develops the ability to control filtration rate and selective re-absorption from the tubules.

There is a clear difference in kidney function between quokkas less than 120 days old and those more than 120 days,[35] which is also the age at which thermoregulation is beginning to develop. The younger animals have higher concentrations of urea in the blood and lower levels in the urine than the older animals (Fig. 2.22), indicating a less efficient clearance of urea through the kidney. Conversely, under the duress of starvation the blood and urine urea levels fell significantly in the younger animals and rose in the older ones, which suggests that the increasing metabolic demands of the older animals were met by catabolizing body proteins, whereas the younger animals reduced their metabolism and protein was conserved. Such a catabolism of protein would suggest that glucocorticoid hormones are being secreted by the adrenal cortex by this

	Pouch young <120 days			Pouch young >120 days		
	Normal	Starved	Starved + Vasopressin	Normal	Starved	Starved + Vasopressin
Urea mg/100 cm^3						
in plasma	51	35		24	66	
in urine	212	143	108	275	471	507
Na$^+$:K$^+$ m.equiv./dm^3						
in plasma	141:5	137:6		150:6	149:7	
in urine	27:28	45:26	96:33	19:38	53:72	70:91
in milk	35:15			30:45		

Fig. 2.22   Kidney function in early and late pouch young of *Setonix brachyurus*. (From Bentley and Shield[35])

age (see p. 156), but this aspect has not been investigated in the pouch young of any marsupial.

The other important function of the kidney is to maintain the ionic concentration of the blood, especially the relative and absolute concentrations of Na^+ and K^+. The normal ratio is about 150 m.equiv. /dm^3 Na^+ :10 m.equiv./dm^3 K^+ in the blood plasma and a ratio nearer unity but favouring K^+ in the urine, which is achieved by the selective re-absorption of Na^+ across the walls of the kidney tubules. Pouch young quokkas of both ages had the normal Na^+ :K^+ ratio in the plasma (Fig. 2.22) but the total electrolyte concentration was lower in the younger group and the proportion of Na^+ in the urine was higher than in the older group, which suggests that selective re-absorption of Na^+ was not taking place in the kidney. It is possible that the plasma ratio was being maintained in these younger animals by a high ratio of Na^+ :K^+ (1·8) in the milk. This high ratio in the milk fell to 0·7, the normal ratio for eutherian milk, after day 120, when the young animal's kidney is assuming the function of regulating plasma Na^+ by selective re-absorption through the kidney tubules. The developing functions of a kidney can be shown by its response to injections of vasopressin, the anti-diuretic hormone of the posterior pituitary gland, which facilitates re-absorption of water across the ascending loop of Henle and the distal convuluted tubule. This can be observed as a concentrating effect on urine; quokkas less than 120 days old showed no such effect after injection of vasopressin, whereas the older animals did (Fig. 2.22). Again this suggests that the posterior pituitary may have become functional by the second half of pouch life, but like with the adrenal gland, this has not yet been investigated.

Immunological competence

Foreign organisms or proteins entering the body of a homeothermic animal generally elicit a defence response which involves the proliferation of special cells by the lymphoid tissues of the body and the production of specific immuno-globulins, known as antibodies; these appear in the γ globulin fraction of the serum and react specifically with the foreign matter or antigen. The ability to mount an antibody reaction develops at about the time of birth in eutherian mammals, coincident with the differentiation of the lymphoid tissue of the thymus. Before this stage the animal or foetus is tolerant to foreign proteins and moreover in subsequent life will remain tolerant to the same proteins if encountered again. During this early period the foetus receives immuno-globulins from the mother,[48] either across the placenta, as in the rabbit and man, or through the milk, as in cattle and horses, in which the first milk, or colostrum, is enriched with maternal immuno-globulins. The immune response is most highly developed in mammals and birds and seems to be an adaptation consequent upon a stable high body temperature, which provides a very favourable environment for pathogenic organisms to grow in.

Adult marsupials, such as the opossum,[268] the brush possum and the red kangaroo have a well developed immune response to antigens, and produce immuno-globulins,[224] so that it is pertinent to find out whether the response develops at a time equivalent to a eutherian neonatus or nearer the time of birth; and whether or not the mother provides immuno-globulin antibodies via the uterus or in the milk. As already noted, γ globulins are at a low level in yolk sac fluid of the tammar and in the serum of young quokkas, although in the latter species they are actively secreted into the milk (p. 84). If the mother does not provide immuno-globulins, the animals' own system must develop at an early stage and this aspect has been most thoroughly studied in the opossum (Fig. 2.23), although work on the quokka[294] and red kangaroo supports it.

The opossum thymus consists of a pair of structures lying near the base of the aortic arch and on the day of birth consists of undifferentiated embryonic cells.[41] Within a day or so the first lymphocytes and the first lymph nodes appear and by day 17 the spleen also contains differentiated lymphoid tissue. Plasma cells and secondary lymph nodes appear at 60 days. The development of lymphoid tissue was grossly affected by removing the thymus at day 7;[180] the number of small and medium lymphocytes was reduced and they failed to appear in the spleen, in which myeloid tissue persisted and increased. This suggests that the thymus plays a role in the origin and maintenance of lymphoid tissue and the suppression of myeloid tissue. Peramelids and dasyurids also have a single thoracic thymus but the brush possum and quokka have a superficial thymus in the neck as well.[124]

The young opossum if injected with a bacterial suspension of *Salmonella typhi*[223] or infected by a dirty wound before day 6[40] will not produce antibody, will not show an inflammatory reaction, and will rapidly succumb to the infection (Fig. 2.23). But after this age, reaction to infection progressively increases and antibody can be detected in progressively higher titres, while wounds rapidly heal. Then again, the young opossum less than 10 days old will accept a skin allograft from its mother[152] and retain it permanently; if challenged with a second maternal graft at the time of weaning it will accept that too, whereas it will reject an allograft from another donor at this later age in 7 days. These experiments show that immunological competence is achieved, and tolerance is lost, in the opossum at a stage of development far in advance of the eutherian embryo, but at a comparable time after birth.

Age in days	Development of Lymphoid tissue[41]	Responses to: Wound infection[40]	Salmonella typhi[223]	Bacteriophage[153]	Skin allograft[152]
1–6	First lymphocytes and lymph nodes by day 3	No inflamation; 90% dead in 2 days	No antibody formation	No response	Accepted permanently. Later isogeneic grafts also accepted
7–10		Intense inflamation, heals in 2 days. Less than 2% dead	Antibody formation at low titres	Low response	
11–22	Lymphoid cells appear in spleen day 17–20				Accepted for 80 days, then rejected
23–60	Plasma cells and secondary lymph nodes		Antibody formation at high titres	Immune response, rejection in 57 h	
61–100					First set rejection in 7–22 days

Fig. 2.23 Development of immunological competence in the opossum, *Didelphis marsupialis*.

The red kangaroo and the quokka, *Setonix brachyurus*, also develop competence at an early stage of pouch life, and in the quokka surgical removal of the superficial thymus has demonstrated that it is the responsible organ.

CONCLUSION

When it emerges from the pouch the young marsupial has developed the anatomical and physiological attributes of a homeostatic animal, and in the quokka, most of these seem to develop after about 120 days. The young eutherian is usually homeostatic at birth or soon after and the age of 120 days in the quokka has been taken to be equivalent, developmentally, to the time of birth in Eutheria.[35] Only immune competence is developed much earlier than in eutherian foetuses of equivalent stage. These comparisons are summarised in Fig. 2.24 by comparing the rabbit, a small eutherian herbivore, with the quokka. This emphasizes the different proportions of time spent in gestation and in lactation but

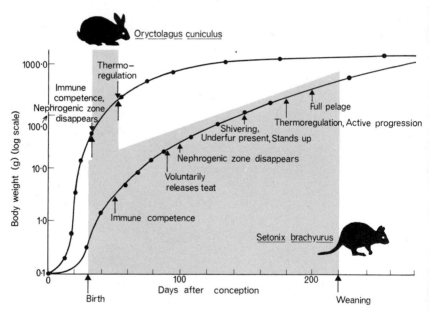

Fig. 2.24 Comparison of the growth and development of an herbivorous marsupial and an herbivorous eutherian of the same adult size. (Data for the rabbit from Brambell,[48] Gersh,[97] Hammond and Marshall,[112] Tyndale-Biscoe and Williams;[283] and for the quokka from Bentley and Shield,[35] Shield,[247] [249] Shield and Woolley,[250] Yadav and Papadimitriou[294])

shows that birth in either kind of mammal is a minor disturbance in the progress of development towards physiological autonomy. Each pattern seems to have advantages. Long intrauterine development allows a fast growth rate during the poikilostatic phase, whereas the pouch affords a longer period of protection in a uniform environment. Whether the slow growth rate of the marsupial is related to its lower metabolic rate (see p. 12) or is an indication that a chorio-allantoic placenta is a more efficient organ than a mammary gland for nourishing an offspring must await more study on more species. What is clear from this simple comparison is that it takes the quokka young four times as long to obtain the weight of the rabbit at birth but they both reach adult size at the same age. However, a female rabbit in 240 days can produce 35 kittens weighing a total of 1700 g, while in the same time, a female quokka has produced one joey, weighing 500 g. Part of the difference is because the young rabbit is independent much younger and provides itself with food, whereas the young quokka is wholly dependent on its mother for much of this time. The subsequent chapters will reveal how effective the marsupial pattern of reproduction is in the economy of the several species studied.

3

Ruminant-like Herbivores—The Macropodidae

The evolution of the mammals and the flowering plants were closely tied to each other. The ancestral Angiosperms were woody plants and flourished in the warm tropical climates of the early Tertiary, as they still do in the tropics today. Herbaceous habit became more frequent from the late Eocene and by the Miocene large areas of herb field and grass land had developed in the cooler, drier climates that then prevailed. The Cretaceous mammals were carnivorous, and the herbivorous forms evolved later. During the Eocene the dominant herbivores were the Perissodactyla (later to give rise to the horses), but they were displaced by the far greater radiation of the Artiodactyla, which have been the dominant herbivores ever since.

GENERAL ADAPTATIONS OF RUMINANT HERBIVORES

Since no mammals can synthesize the enzymes necessary to hydrolyse cellulose, the evolution of herbivorous forms depended on the development of a symbiosis between the mammalian species and bacteria and Protozoa capable of fermenting the cellulose of herbs and grasses. Effective use of bacterial fermentation depends on comminution of the plants by grinding teeth and a slow passage of the material through the alimentary canal, so as to allow time for bacterial action to take place. The fermentation may take place at the hind end of the gut in a large caecum as in the horse and rabbit, or at the front end in a sac derived from the oesophagus or the cardiac portion of the stomach, as in the Ruminantia. Comparisons between horse and cattle show that fibre passes through the horse twice as fast as through the cow or sheep and about one third as much is digested. The advantage of the fore stomach fermentation over the caecal fermentation is not due to differences in the bacterial efficiency, but to the fact that after leaving the fore stomach the fermentation products and bacteria can be digested and absorbed through the length of the small intestine. By reingesting caecal material, including bacteria, into the stomach the rabbits and hares have, indirectly, achieved the advantages of fore stomach fermentation.

Fore stomach fermentation is best developed in the Ruminantia but by no means exclusively by them.[183] It has probably evolved independently in the Tylopoda (camel, llama), the hippopotamus, the sloth, *Bradypus*, and the leaf eating monkey, *Presbytis*, and also in certain marsupials.

The particular features of ruminant digestion, as disclosed by the sheep and ox, are the storage in the reticulo-rumen of freshly eaten plant material; bathed in copious alkaline saliva buffered to a pH of 7–8. In this environment the bacteria hydrolyse the cellulose and other polysaccharides and reduce the monomers by anaerobic glycolysis to volatile fatty acids, especially acetic, propionic and butyric acid. These are absorbed through the rumen wall and provide an energy substrate for the mammal host. As a consequence of bacterial fermentation, glucose is not readily available to the animal and the blood sugar level is characteristically low. As the rumen fluid with its bacteria is sieved into the abomasum it meets a low pH and proteolytic enzymes which hydrolyse the bacterial proteins; these are absorbed along with lipids in the small intestine.

The bacterial symbionts confer additional benefits to the economy of the ruminant; urea, formed by protein catabolism in the mammal, is returned to the rumen via the saliva and across the rumen wall and is re-synthesized into bacterial protein instead of being excreted in the urine. This enables the ruminant to make better use of low protein food and to conserve the water not required for urine formation. In addition, the bacterial flora of the rumen enables new plant environments to be colonized, since they rapidly adapt to, and metabolize, new plant products, some of which might otherwise be toxic to the mammal. Finally, the bacteria can synthesize certain vitamins, notably B_{12}, essential to the ruminant.

Dissections of kangaroos by Everard Home and later Richard Owen led them to recognize, more than 150 years ago, similarities of the stomach to those of the Ruminantia and Owen also noticed that kangaroos at the London Zoo underwent a form of regurgitation and chewing of the cud. Subsequently, in 1876, Schafer and Williams[227] described the differing histology of the parts of the kangaroo stomach and noted their close resemblance to the parts of the ruminant stomach.

Here the matter rested until Moir, Somers and Waring,[184] with the new understanding of ruminant digestion, examined the digestive processes of the quokka, *Setonix brachyurus*. They established that bacterial fermentation of cellulose occurs in the fore stomach and that many other ruminant adaptations occur in this macropod marsupial, and, to judge from the homologies in their stomach anatomy, generally in the Macropodidae.

Students of dentition had independently arrived at the conclusion that the Macropodidae are essentially phalangerid marsupials adapted for grazing, and that the radiation of the family had been from forest-dwelling ancestors, through dwellers in thickets to the predominantly grazing wallabies and kangaroos.[208] Like the Ruminantia, which are the most successful eutherian herbivores, the Macropodidae are the most successful marsupial taxon in the Australian fauna; in both cases it is probable that the mode of nutrition and the metabolic advantages that it confers have been the pre-eminent cause of their respective successes. In this chapter we will look at the macropodid adaptations, comparing them especially with the Ruminantia, and then see how particular species have exploited these in their preferred habitats.

GENERAL ADAPTATIONS OF THE MACROPODIDAE

Masticatory apparatus

The macropod dentition can be derived from the phalangerid pattern by reduction of the front teeth, with a resultant diastema, and by the greater development of the back teeth and associated jaw structure and musculature (Fig. 1.5).

The incisor teeth of macropods, like those of phalangerids, consist of 3 pairs in the upper jaw and a single procumbent pair in the lower jaw. The mandibular symphysis, however, does not fuse, so that some independent lateral movement of the two long incisors is possible. When spread, each one can shear against the three opposing teeth in the upper jaw; when pressed together, both rest between the two upper rows (Fig. 3.3b).

The first two cheek teeth in each jaw of young macropods are a large cutting, or sectorial, premolar (P_2) and a molariform premolar (dP_3). Later on these two teeth are shed (Fig. 3.2) and their place is taken by a single sectorial premolar (P_3). Most Australian authors refer to these premolars as P_3 and dP_4 and the replacement as P_4. Since no marsupial, living or fossil, is known to have four premolars and the basis for this usage has been refuted,[37] I have adopted the notation used by American workers. Behind these teeth the molars erupt so that at all stages of development the molar row is preceded by a sectorial tooth.

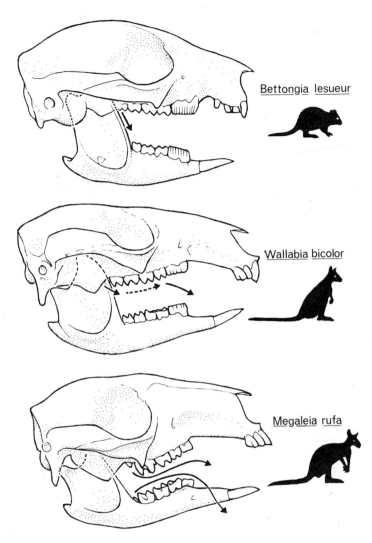

Bettongia lesueur

Wallabia bicolor

Megaleia rufa

Fig. 3.1 Lateral view of the skulls of a rat kangaroo, a wallaby and a kangaroo at the same basi-cranial length. Note the absence of canines and larger size of the diastema in the Macropodinae; the large sectorial premolars and masseteric foramen in the rat kangaroo. The molars in the Potoroinae diminish in size posteriorly and have a fixed position in the jaw, whereas in the macropodinae they are equal sized but move forward in the jaws, being shed progressively from the front. In the kangaroos the premolar also participates in this but not in the wallaby.

The two sub-families of Macropodidae show several divergences in dentition, which may be related to their differing feeding patterns or to difference in size. The Potoroinae (rat kangaroos) are all small animals of about 1–2 kg, whereas the Macropodinae range from this size up to 80 kg. The Potoroinae retain the canine in the upper jaw, the sectorial premolar is a very large tooth and the diastema is relatively short (Fig. 3.1). The first two molar teeth are larger than the back two and all have low crowns with four cusps. They erupt at about the same time and remain in the same position in the jaw throughout life.

By contrast the Macropodinae (wallabies and kangaroos) generally lack canines, and have a long diastema and a relatively small sectorial

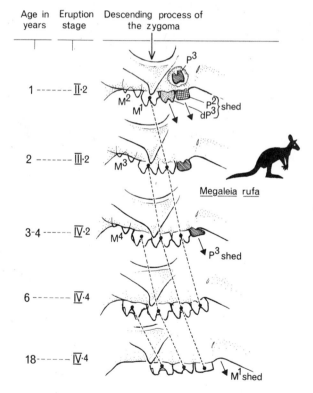

Fig. 3.2 Lateral view of the molar row of a series of known age red kangaroos to show the replacement of the premolars, the sequence of tooth eruption and forward progression past the descending process of the zygoma and the eventual shedding of teeth at the front of the row. (Drawings made from known age specimens in the collection of Division of Wildlife Research, C.S.I.R.O., Canberra)

premolar (Fig. 3.1). The second sectorial (P_3) is retained throughout life in most wallabies but is shed during middle life in the larger kangaroos (Fig. 3.2). The molars are all about the same size, are high crowned (hypsodont) and the cusps are joined in pairs to form transverse ridges or lophs. This development parallels the hypsodonty of the eutherian herbivores and is an adaptation, probably required by the severe attrition of chewing low growing, siliceous plants. As well as this adaptation for grazing the Macropodinae possess another adaptation, which they hold in common with the elephants. The molars erupt sequentially, instead of together, and move forward progressively in the jaw (Fig. 3.2). In the smaller species, such as the hare wallaby, *Lagostrophus fasciatus*,[133] the sectorial moves forward as well so that the diastema becomes shorter in older animals (Fig. 3.1), but in the larger species, such as the red kangaroo, the premolar is shed from the front of the row and its position in the jaw is taken by M1 moving forward and the other erupted molars following. As M1 becomes worn down it too is shed from the front and its place is taken by M2, and so on. Thus at any time an adult kangaroo will have some worn molars and some still erupting and the amount of forward movement can be determined by some reference point on the skull. The descending process at the anterior end of the zygomatic arch is usually used, as this can be felt in living animals (Fig. 3.2). Normally only four molars erupt in each jaw so that very old kangaroos may have only one worn tooth left in each jaw. Thus the life span may be determined in kangaroos, as in elephants, by the durability of their teeth. In one species of wallaby, *Peradorcas concinna*, however, new molars are continuously produced in the back of the jaw.

Molar progression can be a very convenient means of determining the age of kangaroos[244] once the rate of movement has been worked out on animals of known age. It can also be used, without reference to exact age, to subdivide a sample of skulls into classes of increasing age. From this can be learned quite a deal about the structure and dynamics of a population without recourse to live animals; this can be important when surveying island populations of rare species.[133]

Three sets of muscles are involved in mastication by macropods[9] (Fig. 3.3a). The medial pterygoid muscles take origin on the pterygoid bone and insert on the inflected shelf of the dentary; contraction of the muscle on one side draws the dentary medially, while alternate contractions cause a side to side motion. The lateral pterygoid muscle inserts posterior to its origin, close to the articular condyle; its contraction drives the dentary forward.

The masseter muscles oppose the pterygoid muscles. Taking origin along the zygomatic arch, the superficial block inserts all over the posterior half of the outer face of the dentary, while the deeper blocks insert

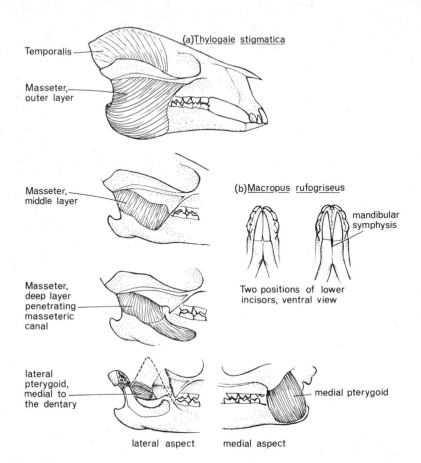

Fig. 3.3 a. Jaw muscles of the pademelon, *Thylogale stigmatica*. Superficial layers and portions of bone displaced to disclose the deeper muscles. (After Abbie[9]). b. Ventral view of a wallaby to show the two positions of the lower incisors in relation to the upper incisors. (After Ride[214])

in the masseteric foramen. This cavity is unique to the Macropodidae and enables the deep masseter to be carried forward to insertion beneath the molar row. It is largest in the Potoroinae where it extends as far as the large sectorial premolar. Abbie[9] suggested that it thereby enabled power to be applied close to the main chewing teeth but Ride[214] considers that it may, with the medial pterygoid muscles on the other side of the articular fulcrum, enable the dentary to be rotated on its long axis. This would allow the premolar of the lower jaw to deliver a greater shearing force against the upper sectorial, as in a well set pair of scissors.

The third important jaw muscles are the temporals which originate on the surface of the cranium and insert on both sides of the ascending process of the dentary. Their contraction draws the jaw backwards as well as aiding in mastication.

By X-ray cinematography of Bennett's wallaby, *Macropus rufogriseus*, Ride[214] described four movements in feeding. As the mouth opens to grasp food the lower jaw moves forward relative to the upper, and the lower incisors separate (Fig. 3.3b). Both movements result from contraction of the pterygoid muscles and enable the lower incisors to bear against the upper incisors. As in ruminants the food is grasped, not cut, and as it comes free the dentary is drawn back and the lower incisors close together, by contraction of the masseter and temporal muscles.

The grass stems are arranged in the diastema by the tongue and lips and the ends are then fed into the molar row, where they are comminuted. This is done by side to side movements of the lower molars across the upper molars, the transverse lophs of the crowns inter-digitating with each other.

After mastication, the food passes to the stomach, from whence it may be regurgitated later and chewed again. However, this process in the Macropodidae is rare[24] and probably not important to digestion as it is in Ruminantia. Like ruminants, though, macropods have large parotid salivary glands,[92] and they produce a copious saliva ($5 \cdot 8$ cm^3/kg/h) with a high concentration of sodium and bicarbonate ions, which act to buffer the fluid at an alkaline pH. The saliva, unlike that of ruminants, contains amylase activity, though much less than human saliva.

The macropod stomach

The macropod stomach is very large and, when full, can weigh 15% of the total body weight, a proportion similar to other ruminant-like animals and considerably more than non-ruminant herbivores, such as the horse. It is an elongated curved bag, deeply sacculated on the outer curvature and smooth on the inner, shorter side. Four regions can be distinguished macroscopically, which correspond to functionally distinct parts[104] (Fig. 3.4). The oesophagus opens into a funnel-shaped region, which extends along the inner curvature as a partly closed spiral groove and the whole is lined with a non-glandular stratified squamous epithelium. This is analogous to the oesophageal groove of the true ruminants whose function is to direct non-fibrous food past the reticulo-rumen to the abomasum. Likewise the macropod spiral groove communicates with a non-sacculated and thick walled chamber, which, with the pyloric region, corresponds to the abomasum of the Ruminantia. It has a full complement of parietal and chief or pepsinogenic cells in its well developed epithelium and consequently has a highly acid pH of 1·8–3·0.

The largest, sacculated region of the stomach is lined by an epithelium of cardiac glands, which secrete mucus but neither proteolytic enzymes

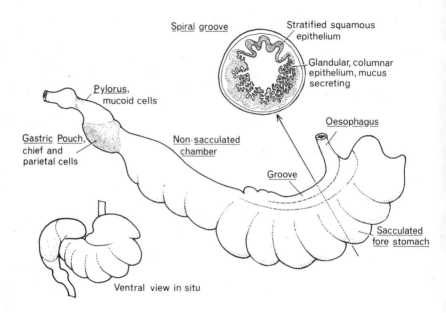

Fig. 3.4   The macropod stomach uncoiled to display the several regions. (After Schaffer and Williams,[227] Griffiths and Barton[104])

nor acid; the pH ranges from 4·6 before feeding to 8·0 after, and is probably maintained in this by the buffering action of the parotid saliva as well as by the mucus secreted by the cardiac glands. This region, which corresponds to the reticulo-rumen of the Ruminantia in these respects, also contains a dense population of cellulytic bacteria (10^{10} per cm^3) and protozoans (10^6 per cm^3).[183]

During pouch life, however, all regions of the stomach of the young red kangaroo show proteolytic activity and a variable but acid pH.[104] Proteolytic activity only becomes restricted to the non-sacculated, fundic, region at the time the young leaves the pouch permanently at 236 days, and when the cardiac region becomes a functional fermentation chamber.

Carbohydrate metabolism

In the adult grazing macropod the fermentation chamber produces volatile fatty acids, ammonia, and gas consisting largely of CO_2 and hydrogen. The volatile fatty acids, as in ruminants, consist largely of acetic acid with lesser concentrations of propionic and n-butyric acid, and are produced at about the same rate (20 μ mole/g/h).[183] Their concentration is very low in the fundic part of the stomach (Fig. 3.5), so that it is presumed that they are absorbed directly across the wall of the cardiac region into the portal blood system.[184] Their concentration in *Setonix*, after fasting or after feeding, was highest in the portal vein draining the stomach,[18] lower in the vena cava after leaving the liver, but variable in the carotid artery; this indicates that the liver has the primary role in regulating the level of volatile fatty acids in the blood. Although this level rises rapidly after feeding, it also falls rapidly, as can be shown by injecting acetic or propionic acid intravenously to unfed quokkas;[18] concentrations were back to pre-injection levels after 30 minutes (Fig. 3.6).

In the case of propionic acid the blood glucose level rose as the fatty acid concentration fell (Fig. 3.6), which indicates that the liver can metabolize propionate to glucose, as does the rabbit. In the fed quokkas mentioned above (Fig. 3.5), the level of blood glucose rose, as also did that of the volatile fatty acids, but the highest concentration of glucose was in the vena cava, not in the portal vein, suggesting again that it was being synthesized in the liver and not absorbed from the gut. Furthermore, the liver of macropods lacks the enzyme glucokinase,[184] which is required to convert glucose to glycogen, so that the role of the liver seems

	Fasted		Fully fed	
	VFA	glucose	VFA	glucose
Cardiac Stomach	13·5	—	63·0	—
Fundic Stomach	2·0	—	6·3	—
Hepatic portal vein	3·5	43·3	13·6	51·2
Thoracic inferior vena cava	1·0	54·6	4·1	72·8
Carotid artery	1·1	48·8	2·4	65·2

Fig. 3.5 Concentrations (mg/100 cm³) of volatile fatty acids (VFA) and glucose in the stomach contents and main blood vessels of the quokka, *Setonix brachyurus*, after 22 h fast and when fully fed. (Stomach values from Moir *et al*[184] and blood values from Barker[18])

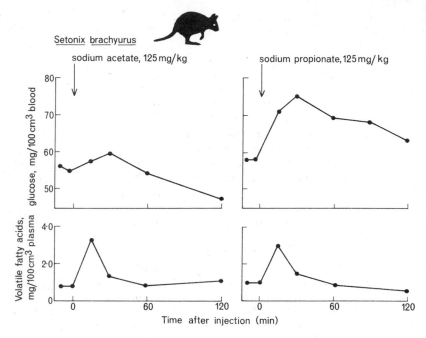

Fig. 3.6 Responses in circulating concentrations of glucose (above) and volatile fatty acids (below) to intravenous injection of sodium acetate and sodium propionate, in the quokka, *Setonix brachyurus*. (After Barker[18])

to be to produce glucose from non-carbohydrate sources (gluconeo-genesis), rather than removing it from the blood. Like the volatile fatty acids, glucose is cleared from the circulation in less than 2 hours (Fig. 3.6) so that the normal low level of 25–78 mg/100 cm³ blood is regained (see also Fig. 2.10 for values in *Macropus eugenii*). This level is slightly higher than that of the Ruminantia, but much lower than in non-ruminants, such as the rabbit.

The Ruminantia, having a normally low blood sugar level, can tolerate severe hypoglycaemia induced by insulin injections, but cannot tolerate abnormally high levels of glucose (hyperglycaemia) as occurs in diabetes mellitus; conversely, non-ruminant herbivores, like the rabbit, rapidly become comatose if the blood sugar falls but can tolerate high levels well. How do the Macropodidae respond to these extremes? The quokka and the red kangaroo, like ruminants, can tolerate quite severe hypoglycaemia without overt effects (Fig. 3.7), whereas kangaroos made diabetic by destroying the Islets of Langerhans with injections of alloxan, develop

	Non-ruminant			Ruminant	
	Oryctolagus cuniculus (rabbit)	Trichosurus vulpecula (possum)	Setonix brachyurus (quokka)	Megaleia rufa (red kangaroo)	Ovis aries (sheep)
Blood glucose (mg/100 cm³)	120	120	78	73–85	65
Response to:					
Insulin (hypoglycaemia)	Coma at 50	Coma at 40	Coma at 20	No effect at 50	No effect
Alloxan (hyperglycaemia)	No effect	No effect until 400	—	Coma at 300	Coma
Glucose injection	Rapid rise and clearance	Very rapid rise (2 min) clearance in 3 h	Rapid rise (15 min) clearance in 1–2 h	—	No rise

Fig. 3.7 Comparison of responses to alterations in blood glucose between ruminant and non-ruminant marsupials and eutherians. (From Griffiths *et al.,*[105] Adams and Bolliger[10] and Barker[12])

a marked hyperglycaemia (300 mg/100 cm^3) and die unless treated with insulin.[105]

Chronic diabetes can be induced by injecting less alloxan and, during the resulting hyperglycaemia, glucose is excreted in the urine (glycosuria). This glucose is supplied to the body by gluconeogenesis from protein in the liver; and consequently protein is therefore unavailable to maintain body weight or growth, so that the animal is in negative nitrogen balance. Insulin injections will redress the effects so that blood glucose falls, glycosuria ceases and a positive nitrogen balance ensues. Both the quokka[18] and the red kangaroo[105] respond to much lower doses of insulin than does the rabbit.

In summary, Macropodidae appear to be intermediate between true ruminant and non-ruminant herbivores. In their dependence on bacterial fermentation, their utilization of volatile fatty acids and their tolerance to low blood glucose they parallel the Ruminantia, but they resemble non-ruminant species in the higher resting levels of blood glucose and in the more rapid response to feeding and starvation. This may be accounted for in part by the faster rate at which food passes through the stomach of macropods, which does not allow the bacteria as long a time to break down cellulose as in the true ruminant.

Nitrogen metabolism

In comparing the metabolic efficiency of the Macropodidae and Ruminantia three factors are important: the efficiency of the micro-organisms as cellulytic agents, the rate of passage through the stomach, and the recycling of metabolic nitrogen and phosphorus to the stomach.

Microbial symbionts

The microbial symbionts in the Macropodidae are taxonomically different from those found in the Ruminantia but appear to be as efficient at digesting cellulose. As mentioned earlier, they utilize the digested cellulose for energy but they also require organic nitrogen for protein synthesis; this may be obtained from the diet or by recycling of urea. Because of their very fast rate of growth, bacteria have a very high ratio of RNA-nitrogen to total nitrogen, which means that they also need a lot of phosphate to synthesize the sugar-phosphate chain of the RNA molecule. Some of this comes from the diet of the ruminant but it has also been observed that cattle and sheep have much higher concentrations of phosphate ion in their saliva than is found in the saliva of non-ruminant mammals, such as horse or man. However, a phosphate concentration equal to that of cattle has been observed in the saliva of one red kangaroo.[92] It has recently been suggested[26] that ruminants recycle phosphate as well as urea and that it is obtained from the bacterial RNA itself when the bacteria are digested in the abomasum and small intestine. The enzyme RNase occurs at far higher concentrations in the pancreas of ruminants than of non-ruminants [26] (Fig. 3.8). It is therefore of particular interest to this idea that the only other vertebrates so far examined that approach the same concentration of RNase in the pancreas are 3 species of the Macropodidae, but not the omnivorous opossum.

| Eutheria | | Marsupialia | |
Species	µg/g	Species	µg/g
Bos taurus	1200	Megaleia rufa	600
Ovis aries	1080	Macropus giganteus	530
Cervus canadensis	550	Macropus eugenii	515
Hippopotamus	62	Didelphis marsupialis	20
Macaca	2		
Homo	1		

Fig. 3.8   The content of RNase in the pancreas of mammals. (From Barnard[26])

Rate of passage

The rate of passage of food through the stomach is a crucial factor in the utilization of cellulose by bacterial fermentation—the slower it travels the more breakdown can occur and hence the less food, or the less nutritious the food, required to maintain a positive energy balance. It has been measured in the quokka and two species of kangaroo, the latter being run concurrently with sheep to give direct comparisons.

The measurement is done by dyeing the food or adding small plastic grains to the diet and recording when the dye or grains first appear in the faeces, and when 90% and 100% has been excreted. These experiments have shown (Fig. 3.9) that the rate of passage in the quokka[59] is faster than in the red kangaroo, grey kangaroo or sheep,[93, 165] but that the latter three species are very similar. More interestingly, all three species show a slower rate of passage when provided with a subsistence diet than when given unlimited food, and under these conditions, a higher proportion of the food is digested (Fig. 3.9).

The sheep in these studies on unlimited diet consumed considerably more than was required to fill their metabolic needs, while the kangaroos did not do so, although young, growing kangaroos ate about 50% more than adults.[91] The effective use of the food can be measured by comparing

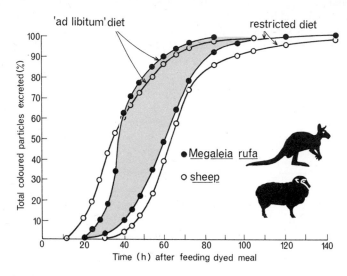

Fig. 3.9 Comparison of the rates of passage of lucerne chaff through sheep and red kangaroos. The rate in both species was slower when the intake was restricted to 600 g and 400 g per day respectively. (From McIntosh[165])

the composition of the food and the faeces in respect of dry matter, fibre, and protein-bound nitrogen. The macropods show very similar values to the sheep (Fig. 3.10), except that sheep digest more crude fibre when fed the same diet. All four species eat less food and digest it less effectively when it has a low protein content than when it has a high one. On the poorer diet, however, they retain more urea in the body.

Urea recycling

The recycling of urea to the rumen, first discovered in sheep, was subsequently demonstrated also in macropods. Amino acids cannot be stored in the body, so that amounts in excess of the anabolic requirements of the body for maintenance, growth or reproduction are deaminated in the liver to ammonia, which is then converted to non-toxic urea. Furthermore, proteins may be converted to glucose in the liver by gluconeogenesis and we have noted that this process is especially important in carbohydrate metabolism in ruminants and macropods. It is possible that some of the resultant ammonia diffuses back across the stomach wall in these species before conversion to urea, but it has been shown that urea certainly passes from the circulation into the fore stomach, where it is synthesized into bacterial protein. As the urea in the

Species	Dry matter intake, g/kg$^{0.75}$/day		DMI digestibility, %		Crude fibre digestibility, %		Crude protein digestibility, %	
	high	low	high	low	high	low	high	low
Setonix brachyurus	44·3	21·4	66	57	42	40	77	66
Megaleia rufa	41·8	24·2	56	48	24	30	73	46
Ovis aries	69·8	43·4	57	52	39	38	73	41
Macropus giganteus	48·1	31·6	54	31	38	27	72	20
Ovis aries	66·5	36·0	58	47	43	43	74	17

Fig. 3.10 Apparent percentage digestibilities of total daily dry matter intake and of the crude fibre and protein components, in two diets; high protein lucerne chaff and low protein oaten chaff or straw. (From Calaby,[59] McIntosh,[165] and Forbes and Tribe[93])

stomach is constantly depleted thereby, the movement from the circulation is probably along a concentration gradient. As well as this, however, urea is a component of the parotid saliva of macropods as it is of sheep, and this is added to the fore stomach contents. The fact that urea is actually utilized in the stomach of macropods has been shown by two experiments. In one it was shown that intravenous injection of urea into euros, *Macropus robustus*, did not result in a higher plasma urea concentration or increased urea excretion in the urine. In a second, more crucial, experiment[53] euros were fed diets supplemented with either additional protein-bound nitrogen in the form of casein, or additional non-protein nitrogen in the form of urea; no consistent differences were found between the two groups of animals in either the amount of nitrogen retained from the food or in the amount of nitrogen excreted as urea.

As well as being recycled to the stomach, urea is excreted in the urine of macropods, and it has been observed in the euro,[54] red kangaroo and tammar[23] that the rate of its excretion in the urine is directly related to the amount of digestible nitrogen in the diet. This suggests that urea may be selectively re-absorbed or concentrated in the kidney in animals on a diet low in nitrogen. Such a function was demonstrated in the camel and sheep and has been shown to exist also in the tammar.[160]

One group of tammars were fed a diet containing sufficient nitrogen to remain in equilibrium, while the other group were fed a deficient nitrogen diet and lost weight during the experiment. Urea concentrations in the plasma and urine remained constant in the first group but both fell markedly in the second group. However, the urine urea was depressed ten times more than the plasma urea, suggesting that urea was being retained by the kidney. And this was shown to be so (Fig. 3.11), for the urea concentration in separate parts of the kidneys rose progressively from cortex to papilla in the first group, while in the second group the concentration fell in the inner medulla and papilla.

Retention of urea by the kidney, as well as being an important adaptation for nitrogen sparing, saves water that would otherwise be required for its excretion; an important adaptation, since water shortage and poor quality herbage are often associated. It has, however, been suggested that the converse may also hold, namely that, under conditions of water shortage, ruminants may recycle more urea as a water sparing adaptation and at the same time achieve a more positive nitrogen balance. The tammar experiences water shortage in some of its habitats (see p. 134) so it was an appropriate species in which to examine this proposition.[25] When fed a low nitrogen diet and with their water restricted, tammars did not excrete less urine but the urea content of it was decreased when compared to tammars on unrestricted water and the same diet. However,

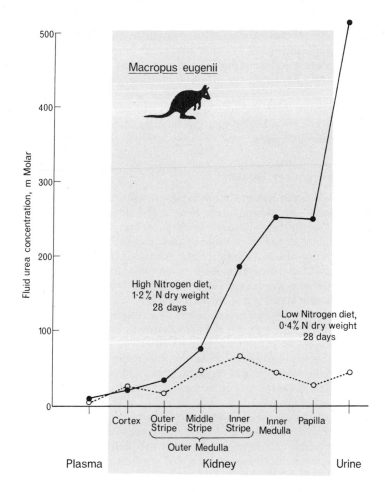

Fig. 3.11 Results of an experiment by Lintern and Barker,[160] to show the difference in urea excretion of tammars held on diets with high and with low concentrations of nitrogen.

the water-restricted tammars passed drier faeces and so reduced water loss by re-absorption from the rectum. Without water the animals ate less food, which would have reduced their nitrogen intake and stimulated re-absorption of urea from the kidneys, as in the previously described experiment (Fig. 3.11); but in that experiment the plasma urea concentration did not change, whereas it fell progressively in the control animals on low nitrogen diet and unrestricted water. Plasma volumes did not change in either group so the water sparing function was not demonstrated in these captive animals. However, in free ranging tammars on Kangaroo Island during the summer Barker[23A] found that urine volume and total urea excretion were diminished and it may be important in this situation where the animals are using much water for thermoregulation. Also the free ranging animals must have greater energy requirements than those in close confinement. With these pieces of evidence we can now compare the overall nitrogen and energy metabolism of tammars and euros with eutherians.

Nitrogen and energy balance

The net daily input to the system is the amount of food eaten less the undigested food passed in faeces, and is the dietary nitrogen intake (Fig. 3.12). For the net nitrogen input, however, faecal nitrogen represents undigested dietary nitrogen and also endogenous metabolic nitrogen resulting from biliary products, dead blood cells and gut epithelium, and in ruminants and macropods, undigested bacterial symbionts from the stomach as well. Metabolic faecal nitrogen can be determined as the faecal nitrogen produced by an animal on a nitrogen free diet; more simply it can be estimated by extrapolating the regression line relating faecal nitrogen and nitrogen content of the diet fed. For the tammar this figure is 411 mg/100 g dry matter intake,[23] which is comparable to the sheep and higher than the non-ruminant pig (Fig. 3.12). The value for the euro,[54] however, is considerably lower than the tammar or sheep, which suggests that it can conserve nitrogen better than can either of the other species.

With this allowance the true nitrogen balance can be determined for adult non-breeding animals under constant conditions, by measuring the daily digestible nitrogen intake and the daily urinary nitrogen output. When tammars and euros were fed on diets ranging from low to high nitrogen content the level of digestible nitrogen intake required to balance urinary loss could be determined. In order to make valid comparisons between animals of different size the three quarter power of body weight or kg $W^{0.75}$ is used. The minimum daily requirements for tammars and euros to remain in nitrogen balance are respectively 248 and 160 mg N/kg $W^{0.75}$/day (Fig. 3.12). These values are considerably lower than comparable values for the sheep, and for the euro dramatically so. Also, the urinary nitrogen excretion of both species on minimum nitrogen diets are much lower than for sheep; in the tammar, as we saw, this is due to renal retention of urea. Ruminants generally have lower minimum urinary nitrogen output than non-ruminants due to the recycling of urea, but the two macropods examined are even more

Species	Weight kg	Dietary N intake mg/kg$^{0.75}$/day	Faecal metabolic N mg/100 g DMI	Digestible N intake mg/kg$^{0.75}$/day	Minimum urinary N mg/kg$^{0.75}$/day	Creatinine excretion mg/kg$^{0.75}$/day
Setonix brachyurus	3·5	234	—	—	—	34·1
Macropus eugenii	3·8	289	411	248	60	30·3
Macropus robustus	12·0	309	270	160	34	24·0
Ovis aries	37–50	414	400–550	360–495	110	62·2
Sus scrofa	24–79	—	100–240	251	131	55·7
Oryctolagus cuniculus	2·0	450	—	—	150	55·3

Fig. 3.12   Nitrogen metabolism in 3 species of Macropodidae and 3 species of Eutheria. (From Barker,[23] Brown and Main,[54] and Fraser and Kinnear[95])

conservative, being exceeded in this only by the camel. This suggests that some other factor, besides ruminant digestion, must account for the low nitrogen requirements and low nitrogen excretion of the macropods. The euro, like the camel, is a desert-living species but this cannot be the explanation for the tammar, which lives in a cool temperate climate and yet has a similar nitrogen metabolism. The other possible factor is that the macropods have a lower metabolic rate (see p. 10) than the eutherian species examined and therefore have lower nitrogen requirements.

It has been shown by Mitchell,[182] for instance, that the ratio of endogenous urinary nitrogen to basal energy, in a wide range of mammals and some birds, is about 0·48 mg N/kJ. It can be seen from Fig. 3.12 that this is so for the pig and the rabbit, but for the sheep and for the two macropods the urinary nitrogen is less than would be expected. This may be due to the recycling of urea, which returns metabolic nitrogen to the body and obscures the relationship. However, there is another end-product of nitrogen metabolism that is not recycled; creatinine is excreted at a rate constant for each species and proportional to body weight and basal metabolism, provided the dietary intake is free of creatinine or its precursor, creatine. In both the euro and the tammar,[95] the daily creatinine excretion on a creatine-creatinine free diet, and at a constant ambient temperature of 25°C, was remarkably constant for each species, despite differences in the protein contents of the diet. The marsupial values are considerably lower than those of a range of Eutheria (Fig. 3.12) and closely reflect the differences in standard energy metabolism referred to in Chapter 1 (Figs. 1.3 and 1.4). Of particular interest, however, is the very low value for the euro compared to the tammar and the quokka, which agrees with its lower minimum nitrogen intake and excretion.

It is unfortunate, in the context of the next part of this chapter, that further comparisons cannot be made between the large kangaroos. This is because the standard energy metabolism has been determined for the red kangaroo but not for the euro, whereas the nitrogen metabolism and creatinine excretion have been determined for the euro but not for the red kangaroo.

In summary then, the highly developed fore stomach fermentation of the Macropodidae is remarkably similar in detail and efficiency to that of the Ruminantia and confers all the same advantages of nitrogen conservation and water conservation, while the lower metabolic rate, common to all marsupials, confers additional advantages on the Macropodidae in lower energy requirements and hence a lower minimum food intake or lower grade food. However, the lower metabolic rate may be an important factor in the slower growth rate noted in Chapter 2, a point of some importance in the realized reproductive rate of marsupial species in competition with eutherian species. By looking at five species of Macropodidae that have been studied in some depth in the last 15 years, we will now see how the macropod organization is fashioned to particular environments.

SPECIFIC EXAMPLES OF MACROPOD ADAPTATION

The desert kangaroos, *Megaleia rufa* and *Macropus robustus*

Distribution

Three kinds of kangaroo are readily recognized, although within each kind many species or sub-species have been described. Grey kangaroos, *Macropus giganteus* and *Macropus fuliginosus*, live in wooded country, which generally requires a yearly rainfall of more than 380 mm (Fig. 1.13), whereas red kangaroos, *Megaleia rufa*, prefer open plains and euros or wallaroos, *Macropus robustus*, live around rocky breakaways.[1] Land clearance has reduced the habitat for grey kangaroos while increasing it for reds so that the 380 mm isohyet is no longer a very clear boundary for their ranges (Fig. 3.13); indeed there are places in western New South Wales where reds, euros and the two species of grey kangaroo can all be found on the same station. Nevertheless, the common range of the red kangaroos and euros is approximately defined by this isohyet and encompasses the vast central area of Australia. Rainfall in this region is very erratic so that an average annual rainfall is of less importance than the distribution of effective rainfall. By this is meant a fall of rain which, after evaporative loss, is still sufficient to induce germination of seed and the growth of ephemeral plants. Since effective rains can occur at any time of the year, though more frequently in the summer, the plants and animals lack well marked seasonal patterns of growth and reproduction, but rather, show adaptations keyed to rain when it falls. This is so of reproduction in the two kinds of desert kangaroo, although in other respects they are each adapted in different ways to the environment. Two factors are paramount in this environment, the shortage of free water and the high summer temperatures. Coupled to these are the absence of shade trees and the low nutritional worth and high salt content of much of the vegetation. Animals in this environment must therefore show special adaptations for controlling their heat load and conserving their water and energy resources. The red kangaroo and the euro each do this in different ways but within each species the physiological adaptations to the desert environment are inter-related; temperature regulation and behaviour, water and electrolyte balance, and diet.

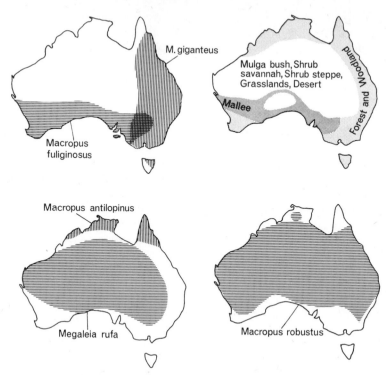

Fig. 3.13 Distribution maps of the large kangaroos and the main types of vegetation in Australia. (From Frith and Calaby[1])

Thermoregulation and water balance

Both species show very good thermoregulatory ability under laboratory conditions of high ambient temperature (40°–44°C) and moderate humidity (30 mm Hg vapour pressure).[221] In one study red kangaroos maintained body temperature at 37·5°C for several hours if water was provided, while in another study of a euro other parameters were measured also.[220] At 40°C the respiratory rate increased from 56 to 271 breaths/minute due to panting so that evaporative water loss also increased, while rectal temperature and pulse remained the same. However, when the humidity was increased to 40 mm the animals showed thermal stress, since evaporative water loss dropped despite increased panting, while the pulse rate doubled and the rectal temperature rose 2°C, approaching the thermal death point of 40°C. From these very limited observations it seems that kangaroos may rely on panting to dissipate heat at high temperatures and that this is satisfactory if the humidity is low and water is freely available. Low humidity is common in their environment but water is often unavailable. Water balance can be maintained, however, by either avoiding high temperatures, as ground dwelling desert animals do, or by efficient concentration of urine.

During the cooler periods of the year euros are active in the daytime but in hot weather they seek the shelter of rock outcrops and caves and they have also been observed to excavate sites under dense mulga bushes, *Acacia aneura*. The microclimate of these sites, especially caves, are not only cooler and more humid than the outside, but also remove the animals from the severe heat load of solar and other radiation.[73] Altogether this behaviour reduces the need for evaporative cooling and hence conserves body water. If water is abundant, however, as after rain or from stock tanks, euros forsake their heat refuges despite high air temperatures. One study[82] in north Western Australia centred around the drinking patterns of euros on a station where the only free water was in stock tanks. Animals coming to drink were caught and marked with collars and it was found that, although some animals came regularly to the tanks in dry weather, others (28%) never did so. Also, among those that did drink, lactating females predominated. After rainfalls of 30 to 40 mm, drinking activity at the tanks diminished markedly. Furthermore, the drinking patterns at different tanks correlated with the presence or absence of rock outcrops in the adjacent country; fewer animals lived around tanks where rocks were absent but they visited the tank more frequently than did the euros living in rocky country.

Taken together, these observations suggest that euros can survive without free water if they have access to cave shelters, but more euros can occupy an area if water is available. This conclusion was supported by the discovery of a sparse population of euros living in rock break-aways 30 miles from free water on the edge of the Canning Basin, Western Australia. No sheep are run there because the ground is too sandy to hold pumped water. Thus it is possible that the provision of stock tanks for sheep has enabled more euros to live in the surrounding country than before they were made.

Euros are sedentary animals and also fairly solitary. Nothing is known about their territorial or social behaviour but it would be most interesting to know whether cave refuges are held as territories, whether this confers dominance, and whether the drinkers at Ealey's[82] tanks were subordinate animals unable to avoid water loss by entering shelters. It seems likely that this is so and not that the non-drinkers are especially adapted physiologically to survive without water, since euros lost 20% of their body weight in 7 days if held in enclosures that lacked rock shelters and free water.[83]

By contrast, red kangaroos are never seen around rocks but remain in the open country and are more mobile and gregarious than euros. In hot weather they seek the partial shade of low bushes, which reduces the

Plate 1   Female euro, *Macropus robustus*, and young animal 1–2 year old. Note the length and loose texture of the pelage. (Stan and Kay Breeden)

Plate 2   Red kangaroos, *Megaleia rufa*, sheltering under a mulga bush. (Edric Slater)

solar radiation to 20% of the open ground value, but the radiant tem-perature still remains as much as 30°C above T_B, while the air temperature is approximately the same as T_B.[73] A large part of the heat load is ameliorated by reducing activity to a minimum and by the dense reflective pelage. A comparison of the pelage of euros and red kan-garoos[71] showed that the pale colour of the red kangaroo reflected a higher proportion of solar radiation than did the darker coat of the euro. It is particularly interesting that the pelage on the hip has significantly higher reflectance than that of the back, since red kangaroos normally rest by lying on their sides. As well, the fur of the red kangaroo is finer and more than three times as dense (62 fibres/mm²) as that of the euro (19·6/mm²) and this greatly increases its insulative properties under wind conditions, both in hot summer and cold winter temperatures. Water lost through evaporative cooling during panting and licking the fore limbs is replenished by selecting better quality food than that of euros; in two different studies it was observed that red kangaroos remained within a few km of persistent food.[193, 265] As well as its greater dependence on water than the euro, the red kangaroo may also be able to conserve water better than the euro; the kidney has a greater concentrating ability (Fig. 3.18) and in one study[74] in New South Wales, where euros and red kangaroos were studied together, the urine of the red kangaroos was consistently, and in summer, significantly more concentrated than that of the euros. This conclusion agreed with previous observations on euros in north Western Australia[83] and red kangaroos in Central Australia.

Nutrition

However, urine concentrations, especially urea, reflect not only the response of the animal to thermal load but also the quality of the diet. It will be recalled that macropods can respond to low nitrogen diet by recycling urea back into the fore stomach, so that lower urine osmolalities in euros may reflect their poorer diet and greater urea recycling. Dependence on rock refuges and a sedentary way of life imposes restrictions on the choice of food to that which grows nearby, so that euros seem to be adapted to live on the poorer quality plants that grow in and around the rocky country. The great ability of euros to conserve nitrogen and to remain in nitrogen balance on low grade feed was noted earlier (Fig. 3.12). This response is well seen in comparisons of the urine urea concentration of euros shot on two stations, one with much less nutritious plants than the other.[83] The euros on the poorer feed had urea concentrations of less than 100 m Molar and were in a lower nutritional state, as judged by weight and haemoglobin, than were the euros of the better property whose urea concentrations were more than 500 m Molar; compare this to the tammar in Fig. 3.11. The nutritional state of the plants greatly improved after rain, and deteriorated during drought when mortality in the euro population was severe. Thus nutrition rather than water may play a major role in controlling population size of euros.

Not so the red kangaroos. No fully comparable study has been done on these animals, but from three separate studies[91, 93, 165] it seems apparent that they cannot utilize low grade feed as efficiently as euros, and are more dependent on high protein feed. This correlates with their well known preference for green feed. On dissection, the stomach contents of red kangaroos are remarkably green even when the vegetation seems dry and brown, and analysis of faeces and stomach contents confirms that the preferred plant species are newly sprouted grasses and ephemeral dicotyledons.[265] It has also been observed that red kangaroos will move considerable distances to feed on a pasture freshened by rain. In Central Australia, Newsome[193] observed by aerial survey during drought and after effective rainfall that the majority of kangaroos during drought stayed within 1·5 km of persistent food. After rain fell, however, many were found 10 km and some up to 28 km from persistent food, as they exploited the new growth of ephemerals and grass shoots. One example of this movement is provided by Newsome.[192] In the drought of 1962 he counted 3900 kangaroos on the study area of 7000 km² while six months later after good rains he counted 4900 animals, indicating an immigration of 1000 animals; in 1966, after a further drought of $3\frac{1}{2}$ years, numbers had fallen to 2800.[196]

Thus we see that these two very successful and widely distributed species of kangaroo have emphasized different aspects of their common macropod heritage in adapting to the desert. The euro keeps near its caves and thereby conserves water, but must perforce be sedentary. It has therefore adapted to the nearby and less nutritious diet, by greater use of urea recycling and a low nitrogen and basal metabolism. Conversely, the red kangaroo has retained mobility, at the cost of more prodigal use of water, for thermal regulation, so that it can travel long distances to exploit the nutritious and succulent plants that bloom after rainfall.

Reproduction

With present knowledge the adaptations for breeding appear to be the same for both species but further study, particularly of contiguous populations, might disclose differences in this as well. Like many other desert animals, the euro and red kangaroo are opportunistic breeders and show no regular seasonal pattern of reproduction, so that breeding must be considered in relation to good conditions, to drought and to the conditions after amelioration of drought (Fig. 3.14). Under favourable conditions nearly all adult females of both species, when examined in the field,[81, 96, 194] are found to have simultaneously one offspring running at heel which suckles an elongated teat from outside the pouch, and a very much smaller offspring attached to another teat inside the pouch, while upon dissection, a diapausing blastocyst can be flushed from one of the two uteri. By the time the pouch young of the red kangaroo is 120 days old the young at heel is fully weaned, but still associates with its mother; when the pouch young is about 200 days old the uterine blastocyst resumes development which it completes about 33 days later, when the older young leaves the pouch (Fig. 3.14a). Post-partum oestrus and ovulation follow and another offspring is usually conceived then. So although it takes a particular offspring nearly 600 days to develop from conception to independence, the female kangaroo can produce an independent offspring every 240 days while good conditions prevail.

As conditions deteriorate, and for red kangaroos this means primarily the quality of available feed,[195] reproductive activity is reduced (Fig.

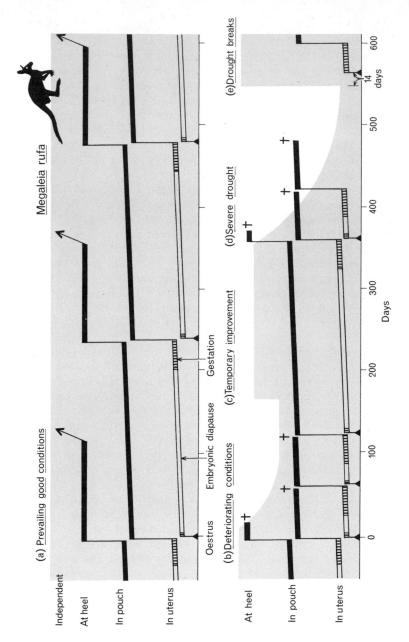

Fig. 3.14 Diagram to illustrate the opportunistic breeding pattern of the red kangaroo, which responds to prevailing drought and rainfall. The same pattern probably holds for the euro, *Macropus robustus*.

3.14b). The first effect is a heavier mortality of the young just emerging from the pouch and changing from milk to an herbivorous diet. It is also the age at which homeostasis has just developed, as we noted in Chapter 2 (p. 89), and water requirements for thermoregulation are probably high. Under these conditions, adult females still carry a pouch young but have another teat that is elongated and dry as evidence of the lost young-at-heel. A factor contributing to this mortality may also be an incipient failure of lactation since Newsome[194] observed that the duration of pouch life was extended under drought and shortened under favourable conditions. Moreover, pouch young during drought were lighter than would be expected from the foot length, used as a measure of age, while, conversely, they were heavier than expected from foot length during good conditions.

If the poor conditions continue, lactation may become inadequate to maintain the pouch young at all. Under these conditions, when females carrying very emaciated pouch young are dissected they are found to have in the uterus an embryo which is advanced in development, it being no longer inhibited by the suckling of its dying sib. When born this young replaces the older and by now moribund young and may complete the first two months of pouch life before itself succumbing to an inadequate supply of milk. Until about this stage the young makes very little nutritional demand upon the mother, so that this replacement of a 2 month old young by a new born can be repeated several times. Its advantage to the species is that the breeding females can immediately take advantage of improved conditions when these occur; lactation will then increase and the small young will continue to grow instead of dying and being replaced (Fig. 3.14c). There is clear evidence from samples of red kangaroos dissected during drought[96, 194] that this adaptation is common and allows breeding to continue well into a drought. It also accounts for the observation that shortly after drought all the females are carrying small pouch young, an observation that gave rise to the erroneous idea that kangaroos have some extraordinary foreknowledge of when a drought will break and anticipate the coming rain.

During very severe drought, when no rain falls for two or three years, the females eventually enter anoestrus so that all breeding ceases and sexual maturity of young animals is deferred (Fig. 3.14d). An adult animal may enter anoestrus while completing a pregnancy[191] so that follicular growth and post-partum oestrus fail to occur, or it may enter anoestrus during lactation, in which case the post-partum corpus luteum remains small and inactive even after lactation ceases.

Effective rain and new plant growth dramatically reverse this condition (Fig. 3.14e). In one study[243] animals were collected two weeks after good rainfall had broken a year-long drought and 65% of the non-lactating females were in or near to oestrus. As it takes about 10 days for follicles to mature, this indicates an immediate response by these animals to the rain, rather than to the new feed the rain would have brought on.

Opportunistic breeding occurs in a wide variety of Australian vertebrates: the amphibian, *Cyclorana platycephalus*, breeds and completes a life cycle in ephemeral pools that come after rain, while many species of native fish come into breeding condition in response to the occurrence of shallow flood waters, as do some waterfowl. Like kangaroos, zebra finches, *Taeniopygia castanotis*, begin breeding activity when rain falls, the young leaving the nest about the time that seed has formed on the plants that grew after the rainfall. Each species uses environmental cues, directly related to improving nutritional and physiological conditions, to initiate reproduction. The only unusual aspect of this compared to the more commonly known ones, such as photoperiod of temperate zone animals, is that the cues do not recur at a regular time each year and so cannot be tied to an annual cycle. For the desert kangaroos in this context the phenomenon of embryonic diapause has obvious selective advantages, so it is not surprising to find that it occurs in almost all females of the population and that the blastocysts regularly survive the whole course of lactation. For seasonally breeding macropods, however, its ecological advantage is less obvious and in some of these it is less prevalent and survival of blastocysts is lower (p. 69). Thus it seems likely that its primary selective advantage is in intra-uterine development, as suggested in Chapter 2, and that the desert kangaroos have exploited a pre-existing phenomenon to the special needs of their own pattern of life, as they have apparently done with the other macropodid features of digestion, kidney function and metabolism.

The forest kangaroos, *Macropus giganteus* and *Macropus fuliginosus*

The two species of grey kangaroo live in those parts of Australia where seasonal changes of climate are more regular and rainfall more uniform and predictable (Fig. 3.13). This is reflected in their biology, especially reproduction, but at present very little is known about the physiology of the grey kangaroos so that it is not possible to relate field observations to physiological requirements or responses. The only two field studies on grey kangaroos have been conducted in southern Queensland, where *Macropus giganteus* occurs, together with red kangaroos and sheep, and all three species have been studied in parallel.

The faecal pellets of the three can be distinguished, so that patterns of their distributions in different habitats can provide indirect evidence of distribution and preferences.[61] Food preferences of the three species were determined directly by examination of buccal and stomach contents of freshly shot specimens at different times of the year.

In Caughley's study[61] the distribution and density of faecal pellets from grey kangaroos correlated with the density of the vegetation, being more abundant in the denser vegetation and almost absent from the open, treeless, grasslands. This agreed with sight traverses through the same habitats when grey kangaroos were more commonly found in the woodlands. Conversely, the red kangaroos were seldom encountered in the denser vegetation, but frequented the open woodlands and grassland; faecal pellets of red kangaroos, however, did not correlate with any particular habitat as did those of the greys.

Notwithstanding the preference of grey kangaroos for dense cover they never ate the foliage of either the mulga, *Acacia aneura*, or berrigan, *Eremophila longifolia*,[103] which are the main woody species, but their diet comprised predominantly (64%–79%) species of grass, such as spinifex, *Triodia mitchelli*, growing in the woodlands. The predominant species of the open country, mitchell grass, *Astrebla pectinata*, was absent from their diet but comprised a major part of the grassy diet of the red kangaroos. Like the grey kangaroos the red kangaroos eschewed the mulga and berrigan, but unlike the greys, more than half their diet consisted of small dicotyledonous plants, as we noted previously in comparing them with euros.

The sheep on the same pastures had dietary preferences distinct from both species of kangaroo. They browsed mulga and berrigan in all seasons, and like the red kangaroos, selected succulent dicotyledons. However, the particular species selected differed; sheep chose species of Malvaceae not eaten by red kangaroos and the red kangaroos selected *Portulaca olevacea* which was never found in the sheep stomachs. In their browsing of mulga, the sheep resemble the euros. Thus, in favourable conditions the three species were each selecting different species in the pasture, but as the pasture deteriorated, due to lack of rain and an influx of additional sheep, all three species were obliged to eat more of the perennial grasses, such as *Triodia*, and showed less diversity in their diet. Analogy can be drawn in this to the remarkable association of the great mixed herds of herbivores living on the east African grasslands, where it has likewise been shown that each species has particular food preferences, which do not overlap with those of the other species.

Unlike the desert kangaroos, the grey kangaroos are not opportunistic breeders, but produce young only in the spring and early summer. Growth of pouch young is slower than in red kangaroos, so that the young leaves the pouch after about 280 days and is still running at heel when its mother next breeds. As mentioned in the previous chapter, post-partum ovulation does not occur and grey kangaroos generally do not carry diapausing blastocysts, so that the whole pattern of breeding in these two species is different from the desert kangaroos.[64] However, a female grey kangaroo can undergo oestrus and conceive during the course of lactation and, when this happens, the blastocyst diapauses until the conclusion of pouch life of its older sibling; thus the mechanism is there, but it does not occur as a normal feature of breeding as in the desert species. This is further support for the idea that a pre-existing mechanism has been taken over as an additional adaptation to desert life.

The island wallabies, *Setonix brachyurus* and *Macropus eugenii*

Distribution

The smaller species of Macropodidae, designated wallabies, are widely distributed throughout Australia and New Guinea and occupy many different habitats. European settlement and the alteration of the land has affected these forms more adversely than it has the kangaroos, and many species are now rare or extinct on the mainland (see Fig. 7.2). However, many of the islands on the continental shelf carry populations of one or more species of wallaby and some of these species have been the objects of considerable research. Island populations are studied not only because of their convenience, in the face of dwindling mainland populations, but because they invite intriguing questions about their own origins and the state of the adjacent mainland at the time of separation, and also about the way the species has adapted to the very local conditions of a particular island.

Two island species have received most attention in recent years, the quokka, *Setonix brachyurus*, a rabbit-sized wallaby weighing about 3–4 kg, and the tammar, *Macropus eugenii*, a somewhat larger animal of 3–5 kg. The quokka is endemic to south Western Australia and its relationships to other macropods are uncertain, whereas the tammar occurs in south Western Australia, probably as far north as Geraldton (29°S), as well as in South Australia. In eastern Australia a very closely related species, the parma wallaby, *Macropus parma*, occupied the same habitat, but is now extremely rare, as is the tammar in South Australia. Rock wallabies, *Petrogale*, the hare wallabies, *Lagorchestes* and *Lagostrophus*, and the rat kangaroos, *Bettongia*, also occurred on the mainland of South and Western Australia and are found on some islands. At first glance the distribution of species on the occupied islands of the south and west coasts (Fig. 3.15) appears to be quite haphazard but closer analysis of the islands reveals some interesting correlates.[167]

The age of each island since separation from the mainland can be estimated from knowledge of the past sea levels by measuring the depth

Plate 3 Quokka, *Setonix brachyurus*, from south Western Australia. (Stan and Kay Breeden)

Plate 4 Tammar, *Macropus eugenii*, in scrub on Kangaroo I. (Edric Slater)

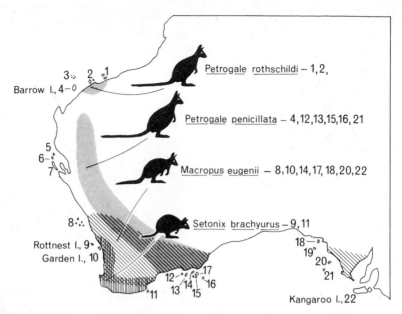

Fig. 3.15 The mainland distribution of 4 species of wallaby in 1900 and the offshore islands on which they severally occur. Four other species, with a more widespread distribution also occur on some of the islands: *Bettongia lesueur*—3, 5, 6, 7, 19; *Lagostrophus fasciatus*—5, 6, 7; *Lagorchestes hirsutus*—5, 6, 7; *Lagorchestes conspicillatus*—3, 4. Kangaroos, whose mainland distributions are shown in Fig. 3.13, occur on the two largest islands: *Macropus robustus*—4; *Macropus fuliginosus*—22. (From Shortridge,[253] and Main[167])

of water in the shallowest place between the island and the coastline. By the south western coast, rock wallabies occur only on islands more than 10 500 years old, whereas with one exception, tammars and quokkas occur on islands younger than this. The exception is the Houtman's Abrolhos group (No. 8) off the coast from Geraldton, which are 11 500 years old, but have no rocky habitat suitable for *Petrogale*. Today on the mainland *Petrogale* tends to occupy a drier environment than do the other two species, which may indicate that the climate of south Western Australia was drier 11 000 years ago than it is today, so that quokkas and tammars were not available to occupy the older islands at separation, and arrived subsequently but earlier than 9000 years ago.

Conversely, it could be argued that several species occupied each island at the time of separation and that species absent today became extinct through inadequacies of the habitat or inter-specific competition. On a restricted area it is possible that species, whose habitat requirements on the mainland did not infringe each other, would be driven into direct competition. A clue that this may be so is found by relating the habitable areas of the several islands to the species. There are many islands that carry no wallabies and the smallest inhabited islands are 12 and 15, each about 1 km²; these carry *Petrogale*. The smallest island with tammars is 14, which is 2·9 km² and this may be near the limit, since tammars have died out, apparently from natural causes, on North Island, Abrolhos, which is 3·6 km². Islands of 10 to 15 km² have only one species of wallaby each, but Bernier Island and Dorre Island (5 and 6) are each 49 km² and carry large populations of three species, as does Barrow Island of 200 km². It is also perhaps significant that Barrow Island and Kangaroo Island in South Australia are the only two islands to carry a species of kangaroo as well as wallabies. It is unlikely that fossil remains will be uncovered on any of the smaller islands that will shed light on past interspecific encounters, but the multi-species islands give an indication of how several species co-exist.

Bernier and Dorre islands are unique in this respect because scientific parties have visited them during a span of nearly 300 years[217] and yet they have been barely affected by human interference; aborigines did not reach them and sheep grazing has never been practised because of lack of water. The three macropod species on both islands are a rat kangaroo and two hare wallabies. The rat kangaroo, *Bettongia lesueur*, lives in burrows in the cliffs along the edge of the islands and scavenges on the shore line; the Western hare wallaby, *Lagorchestes hirsutus*, is a solitary animal which lives in the spinifex tussocks covering the sand dunes; the banded hare wallaby, *Lagostrophus fasciatus*, lives gregariously in the low divaricating shrubs, which cover most of the islands. Studies[263] made on a colony of *Bettongia* derived from Bernier Island disclosed that territorial behaviour was well developed, especially between females, who would vigorously defend their own burrows. On the islands the *Bettongia* populations seem to have been stable throughout their known history. Not so the hare wallabies; the evidence of successive narratives suggests that the relative abundance of the two species has altered, as well as the absolute numbers. In 1896 and 1899 both species were equally abundant but by 1906 while *Lagorchestes* was moderately plentiful, *Lagostrophus* was found in swarms, many of the animals being emaciated and dying. In 1910 the next visitor recorded that none of the wallabies were at all numerous and no specimen of *Lagorchestes* was collected. In 1959 *Lagostrophus* was again abundant and *Lagorchestes* rare, while four years later *Lagorchestes* was almost as common as *Lagostrophus*. These uncontrolled and somewhat subjective observations suggest that the populations may undergo fluctuations in size, which on a smaller island might lead eventually to the extinction of the less resilient species.

Between Rottnest Island and Garden Island (Fig. 3.15) some such interspecies competition leading to extinction probably occurred, for both islands are derived from an arc of consolidated limestone dunes and yet Rottnest Island has a population of quokkas and Garden Island a population of tammars. Both species have been investigated in considerable detail and the results shed light on their respective adaptations to island life.

The preferred habitat of quokkas on the mainland of Western Australia is the dense vegetation surrounding permanent swamps, which provides them with fresh water, deep, cool shade and fairly nutritious food. In this habitat they breed throughout the year. Tammars prefer the drier regions of the same parkland so that the two species do not normally overlap.

Rottnest Island

Rottnest Island has the same climate as the mainland, where 93% of the annual precipitation falls in the winter and the summer is very dry and very hot (Fig. 3.16) with day temperatures reaching 38°C. On Rottnest the winter rainfall of 680 mm is not retained, but seeps away through the porous soil, and the few lakes are saline. The vegetation reflects the severe climatic fluctuation; it is predominantly tussock grassland with scattered low shrubs, which grows during the winter, and progressively declines in water and nitrogen content through the summer, to a low point in March. During the summer the only fresh water occurs at a few seepages or soaks around the salt lakes and on some beaches, but the West end is quite devoid of water. This is a promontory of 0.5 km² joined to the rest of the island by a narrow isthmus of 200 m. During 10 years no quokka marked on the West end has ever been recovered on the other side of the isthmus so that comparisons between the quokkas of West end and the Lakes area provide a means of determining the role of free water in the economy of the species.

Quokkas in the Lakes area are widely dispersed during the winter. When the temperature rises in November they start to converge each night from distances up to 2 km on the fresh water soaks around the lakes, unless rain falls, when these are temporarily forsaken. Since the soaks are small, large numbers of quokkas may congregate at each one and in a 3 year study Dunnet[79] was able to catch and mark respectively 1200 and 1000 at two separate soaks. Many of these were subsequently recaught at the soaks or by hand net in the surrounding country. This disclosed the interesting fact that each soak was used exclusively by the group of quokkas from the surrounding area. Only 3 quokkas marked at one soak were ever recaptured at the other soak; furthermore, young quokkas stayed in their natal territory and did not disperse as adults.

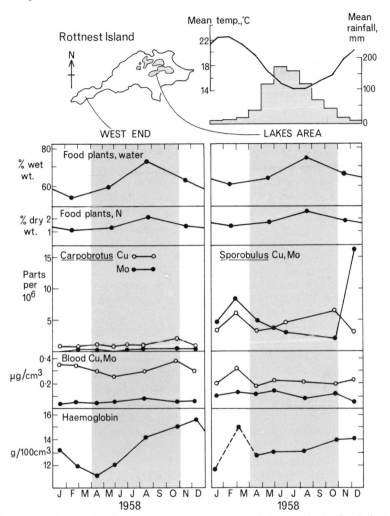

Fig. 3.16 Composite diagram to show how seasonal changes in the food plants on Rottnest Island are reflected in the concentrations of trace elements and haemoglobin in the blood of quokkas, *Setonix bachyurus*, from two separate populations. One population lives on West end, which entirely lacks fresh water, and the other lives around the Lakes area, where fresh water seeps occur through summer. (After Storr,[264] and Barker[20, 21])

It seems clear that the soaks are important to these animals during the dry season but on the West end no such water is available. On West end Holsworth[128] showed that each adult quokka occupied a home range of about 2·4 ha and that the home ranges of a group of 25–150 adults overlap to form a group territory; the whole of the West end could be subdivided into 15 such group territories. Very few quokkas crossed group boundaries, which often coincided with topographical features, even when severe depletion of an adjacent group occurred, as in the harsh summer of 1961–2.

This restriction is particularly important as it shows that the animals are entirely dependent on the resources of their own territory and do not move to more favourable areas, as does the red kangaroo. In this respect quokkas resemble the sedentary euros and, also like the euros, quokkas on Rottnest show marked seasonal changes in condition.

By the end of summer, when the vegetation has very severely deteriorated, a concurrent decline is seen in the quokkas, many of whom are emaciated and about to die. Many quokka skulls lie about on the island and a sample of 600 was analysed for age at death, from which it was concluded that a majority of deaths occur each year at the end of summer, particularly of those born one year before. The lowered condition of quokkas in summer is reflected in a widespread anaemia. In well fed captive quokkas the haemoglobin level is 16 g/100 cm^3 blood, which was also the level in animals collected in the winter from Rottnest. In late summer this can fall to 11 g/100 cm^3 (Fig. 3.16), the West end quokkas showing much more fluctuation from summer to winter than the animals near the Lakes, which had access to free water. During the ten year study on the West end the population declined severely in 1953–4, 1957–8 and 1961–2; these were the only summers when less than 30 mm rain fell in November through March, the drought extended into April, and on more than 20 days the temperature exceeded 29°C.

It seems evident that the harsh summer conditions must exercise the major control upon the quokka population, more especially on West end, and much of the work has been directed to finding out how this comes about. Is water the limiting factor, or diet, or more subtle deficiencies such as the trace elements cobalt and copper?

It was known from early failures to run sheep on Rottnest that the island soils are deficient in cobalt. Ruminants, such as the sheep, have a poor ability to absorb vitamin B_{12} and they depend upon the rumenal micro-organisms to synthesize it for them. If cobalt is deficient in the diet this adversely affects the rumenal flora and its ability to synthesize B_{12}. After the ruminant nature of its digestion had been discovered it was pertinent to ask how the quokka could survive on Rottnest where sheep could not, and it was postulated that the seasonal anaemia, with associated low B_{12} levels, might be due to a cobalt deficiency. However, when quokkas were fed on a diet containing less cobalt than Rottnest plants, their liver cobalt levels fell but they did not develop the typical symptoms of cobalt-deficient sheep; nor did their B_{12} levels fall as low as those of animals on Rottnest. The quokka is thus less dependent on its rumenal flora than the sheep for its B_{12} requirements, or is more efficient at incorporating cobalt into B_{12}; the seasonal anaemia must therefore have other causes.

Ruminants are also sensitive to copper deficiency, which may result from a copper deficient soil. It may also result from a combination of normal copper levels and high molybdenum and sulphate levels, in which the interactions of the three ions reduces the availability of copper and then this in turn adversely affects the haemoglobin concentration of the blood.

Barker[21] showed that the same ionic interaction between copper and molybdenum occurs in quokkas and he found that the concentrations of these elements in the Rottnest plants, eaten by quokkas, varies seasonally (Fig. 3.16), and is at all times less than the minimum requirements for sheep. The copper levels were higher in favoured food plants, such as *Sporobolus*, from the Lakes area than in plants from the West end, but so were the molybdenum and sulphate concentrations in the Lakes area plants (Fig. 3.16). In consequence, quokkas from the Lakes area had lower blood copper levels than the West end animals, although higher molybdenum levels. The blood copper levels correlated with seasonal changes in the haemoglobin concentration of the blood and to a lesser extent with the copper levels in the plants eaten. Quokkas from the West end, on the other hand, had higher and more uniform blood copper levels throughout the year, despite wider fluctuations in haemoglobin concentration. At the end of winter their high haemoglobin concentration may have been aided by the high blood copper, but the same high copper level through the summer did not prevent a drastic fall in haemoglobin concentration at that time. It thus appears that copper deficiency, like cobalt deficiency, may be associated with the seasonal summer anaemia but is not its primary cause.

The quokka has excellent thermoregulatory ability and can maintain a deep body temperature of $37 \cdot 5°C$ against an ambient temperature of $44°C$[28] (Fig. 3.17), which considerably exceeds the maximum temperature experienced on Rottnest. It achieves this, as does the tammar,[76] by panting and by profuse sweating from the paws and fore arms. It licks its arms and appears to salivate too and it was at first thought that the evaporative cooling of the saliva was the important factor in heat regulation, but this was shown not to be so[34] by placing a wide collar around the neck. This prevented the animals from licking their arms, but these still became wet and the body temperature was still kept below ambient. Depriving quokkas of water under hot conditions, however, quickly induced debility, whereas they could survive starvation much

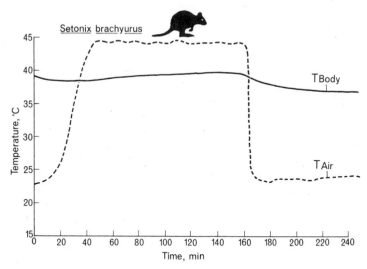

Fig. 3.17 The effect of high temperature on the body temperature of the adult quokka, *Setonix brachyurus*. (From Bartholomew[28])

better; indeed, when dehydrated, quokkas ate less. Evaporative water loss also declined during dehydration, probably because of limitations of the kidney's capacity to concentrate urine and thus to provide water for thermoregulation. In the tammar, evaporative cooling by panting is the main method of dissipating excess heat.[76]

The urine flow of quokkas caught on Rottnest during winter averaged 25 cm³/h and had a mean concentration of 456 mOsm/cm³,[33] which was similar to values of experimental animals given food and water *ad lib.* By contrast the urine flow of animals caught in summer averaged 3 cm³/h or less and the concentration was 1163 mOsm/cm³. The highest urine concentration of a wild caught quokka was 1931 mOsm/cm³, which exceeded the values in experimental animals deprived of water for 2 days at 21°C.

Dehydrated animals would drink salt solutions up to 3% but were only able to keep in water balance on 2·5% saline, by drinking large volumes and excreting a very concentrated urine. The intake of one animal on this regime was 8·0 g NaCl on the fourth day and output in the urine 8·78 g; the next day it took in 14·28 g and excreted 11·33 g, so it was no longer able to discharge the salt load. From these studies it appears that the quokka kidney can produce a urine of about 2000 mOsm/cm³, which is nearly 7 times the plasma osmotic concentration

	Electrolyte, m Eq/dm³		Urea, m Molar		Concentration, m Osm/dm³		Concentrating ability, U/P	Medullary thickness ratio	Refs.
	plasma	urine	plasma	urine	plasma	urine			
Marsupialia:									
Megaleia rufa	150	880		(1200)	312	2780	8·8	5·8	74
Macropus robustus	167	990		(650)	310	2430	7·7		74, 83
Setonix brachyurus	160	693		286		2188	6·9	5·5	33, 35, 57
Macropus eugenii			6	574	316	(2160)	6·9	5·9–6·8	146
Trichosurus vulpecula	151	173	3–6	(575)		1504	4·7		209
Dasycercus cristicauda		645		2450		(4000)	12·7		228
Didelphis marsupialis	151		4–7	212–529	318	1497	4·7	5·7	137, 205, 206
Eutheria:									
Sus scrofa	155		10–14	3800		1080	3·5	1·6	78
Dipodomys merriami				6000		6000	15	8·5	78
Canis familiaris				1000		2425	7·4	4·3	78
Homo sapiens	148	370	4	800		1160	4·0	3·0	78

Fig. 3.18 Comparison of kidney function in 7 marsupials and 5 eutherians. ( ) enclose calculated values.

(Fig. 3.18) and it is not able to utilize sea water or less concentrated salt water to maintain water balance. In the Lakes area, water balance can be maintained at the soaks but on West end the vegetation must provide most of the quokkas' water needs.

During the same years as these other studies were being done Storr[264] examined the diet of quokkas on West end and the Lakes area. He first determined the daily food intake by measuring the rate of passage out of the stomach in samples of quokkas at different seasons of the year. He found this to be remarkably constant throughout the year, suggesting that quokkas monitor bulk rather than quality of the food eaten. At the same time he measured the relative abundance of different plant species in fresh faeces of quokkas held overnight in sacks, and analysed the water and nitrogen content of these plant species. From this he found that the Lakes area quokkas ate several grasses and one species of succulent, *Arthrocnemum halocnemoides*, during the winter and spring, but in high summer the succulent composed 70% of their diet. On West end a different species of succulent, *Rhagodia baccata*, several grass species and the shrub, *Acacia rostellifera*, made up the bulk of the diet in all seasons, except on the far western Cape Vlaming, which is covered by the very succulent *Carpobrotus aequilaterus*, and where this species predominated in the diet throughout the year. The water and nitrogen content of these diets varied seasonally (Fig. 3.16), as a result of changes in the component species and because of the shift in preference for succulents in summer. Since the daily food intake did not vary through the year, it is clear that the amount of nitrogen and water ingested must have varied with the season; this is shown to have been so in Fig. 3.19 for the Lakes area, West end and Cape Vlaming.

The quokka requires a daily N intake of about 0·6 g to remain in nitrogen balance (Fig. 3.12); in the winter months they get more than this, but in summer they get less. The *Carpobrotus* at Cape Vlaming provides plenty of water even in summer but its nutritive quality is so low that the quokkas living on this part are permanently undernourished. To a lesser extent this applies to the rest of the West end; impelled to obtain water from the diet, the quokkas in summer turn to the succulent *Rhagodia*, but its low nutritive value does not provide sufficient nitrogen to keep the animal in balance. The quokkas of the Lakes area, however, can do marginally better in summer because the seepage waters in their territory allows them to eat other plants as well as succulents.

The seasonal debility of Rottnest quokkas compared to mainland quokkas appears, therefore, to be due primarily to insufficient water for metabolic and thermoregulatory needs, so that less nutritious but succulent plants are eaten, thus compounding the debilitating effects of water deprivation with nitrogen shortage as well. The deficiencies of

	Nitrogen, g/day				Water, cm³/day			
	Feb.	May	Aug.	Nov.	Feb.	May	Aug.	Nov.
West end	0·5	0·5	1·0	0·8	110	120	190	160
Cape Vlaming	<0·4	<0·3	—	<0·6	<270	<230	—	<270
Lakes area	0·6	0·5	1·1	0·8	180	110	360	180

Fig. 3.19 Daily intake of nitrogen and water from herbage, of adult male quokkas, *Setonix brachyurus*, at different seasons and localities on Rottnest Island. (From Storr[264])

cobalt and copper are not in themselves limiting factors in the quokka's economy, as they are for sheep, but probably exacerbate the malnutrition and water deprivation.

In one respect, however, the copper deficiency may play an important role on Rottnest. Unlike on the mainland the breeding season on Rottnest is brief.[248] Females come into oestrus in late summer; in a mild year in January and in a harsh year not until March. The young is carried in the pouch until August when it emerges and continues to suckle until October. Although most females mate post-partum and carry a diapausing blastocyst, the proportion of these that will resume development steadily declines (see p. 69), so that very few indeed are born after emergence of the older offspring, the females generally entering anoestrus. If brought to the mainland, Rottnest females do not enter anoestrus and their breeding pattern resembles that of mainland quokkas. Chronic copper deficiency leads to anoestrus in rats, so it is particularly interesting that Barker[22] observed a significantly lower blood copper level and associated lower haemoglobin levels in female than in male quokkas during October, but at no other time of the year. He ascribed this to lactation and, observing that copper levels in the milk were 3 times as high as in blood, concluded that provision of copper stores for the suckling were debilitating the mother. More subtly, the lowered copper levels may be the factor causing the females to become anoestrus.

Water is thus the primary controlling factor for the quokka on Rottnest. If quokkas could concentrate urine more effectively they could perhaps use the water of the lakes or the sea to maintain water balance, but Bentley's experiments demonstrated that they cannot do this. There is no free water on Garden Island or on the Abrolhos Islands where tammars occur, so it has been of considerable interest to discover that tammars have a far greater renal concentrating ability (Fig. 3.18) than quokkas and that they can maintain weight on sea water and low protein diet for up to 30 days.[146] Very few mammals have such a great ability to concentrate urine and thrive on sea water and all those that do are smaller than tammars and most are desert dwellers whose renal capacity is an offshoot of adaptation for water conservation, as we shall see in Chapter 5. Tammars have been observed to drink sea water from the beach on Garden Island and on the Abrolhos islands (Fig. 3.15, 8) after hot weather, so it is likely that sea water plays a real part in the water economy of the species. This ability is reflected in the structure of the kidney. The relative width of the medulla is greater in species that can concentrate urine effectively, presumably because of the increased length of Henle's loop and a wider counter current exchange system. The medullary index for the quokka is 5·1–5·8, for the Garden Island tammar 4·98–6·82 and for the Abrolhos island tammar 6·40–7·90 (Fig. 3.18).

Returning to the postulated competition between quokkas and tammars on Garden Island and Rottnest, the lack of water on Garden Island would certainly have favoured the tammar over the quokka. It is more difficult to envisage the conditions on Rottnest Island that would have favoured the quokka over the tammar unless it be that the vegetation on Rottnest, especially around the lakes, was originally more like the preferred mainland habitat of quokka than of tammar; if the lakes contained sweet water at the time of separation this would have provided all the conditions optimal for quokkas.

CONCLUSION

This chapter began with the notion that macropods are pre-eminent among marsupials because of the exploitation of fore stomach fermentation, in which they closely parallel the Ruminantia among Eutheria. The several studies that we have surveyed show that the comparison is indeed a close one, and convergences of anatomy, of physiology and of biochemistry are to be found. If the Macropodidae are not quite as efficient at utilizing herbage as the Ruminantia, they are also not so sensitive to deficiencies of trace elements in the soil and their lower metabolic rate enables them to maintain health on a lower intake of food. In assessing the efficiency or adequacy of a particular type of animal organization it is instructive to see how it copes at the limits of its environmental extremes. The adaptations of the desert kangaroos, and the adaptations of wallabies isolated on deteriorating island habitats, show how the macropod organization can be adapted to cope with the competing requirements of thermal load, scarce water and impoverished feed.

4

Non-ruminant Herbivores—the Phalangeridae

Two marsupials, more than all other species, have been examined by a variety of workers for a variety of reasons. The North American opossum, *Didelphis marsupialis*, and the Australian brush possum, *Trichosurus vulpecula*, have been used because they are the most readily available marsupial in their respective countries. Unfortunately, some confusion is caused by them sharing the same trivial name.

Apart from these two aspects, they are not alike at all, the opossum being an omnivorous polyprotodont marsupial, while the brush possum is an herbivorous diprotodont. Because of the large amount of work done on it the brush possum will serve as the major example of the non-ruminant, or non-macropod herbivore, and other phalangerids will be compared to it where information on them is available. The pigmy possums, which should be included here on taxonomic grounds, will be considered in Chapter 5 with other marsupials of small size.

PHYSIOLOGY OF TRICHOSURUS VULPECULA

Genetic polymorphism

The brush possum is an animal of the forest and woodland and is common in the winter rainfall areas of Australia, as well as on Tasmania and Kangaroo Island. However, it also occurs in the drier regions along water courses, where it lives in the river gums, so that its distribution is extensive. It does not occur in New Guinea or the northern islands, where

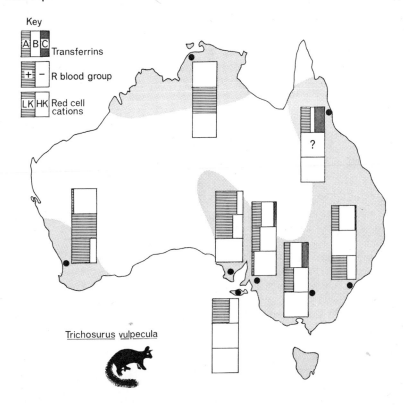

Fig. 4.1 Distribution of the brush possum in Australia, with the frequencies of three genetically determined properties of the blood from 8 local populations superimposed. The frequencies of the three genes determining transferrin proteins are shown in the top cells, while the middle and lower cells represent frequencies of the phenotypic character, not the gene frequencies, of the R blood group and the relative concentration of cations in the red cells. All the frequencies are based on samples of more than 20 animals except for the R phenotype in Perth (12) and the two northern localities of 7 and 5 animals respectively. (From unpublished Ph.D. thesis of R. M. Hope, 'Genetic variation in natural and laboratory populations of *Trichosurus vulpecula* and *Sminthopsis crassicaudata*', University of Adelaide, 1970, by permission)

its niche is occupied by the cuscus, *Phalanger orientalis*, but it has been introduced to New Zealand, where it is now abundant.

Across the continent the species shows marked differences in body size from about 1 kg in northern Australia to 4 kg in Tasmania, and from pale grey pelage in the north to dark brown in the south. Likewise some properties of the blood show regional differences and the genetic basis of some of these polymorphisms have recently been examined by R. M. Hope. Three common alleles determine the transferrin protein polymorphism and the relative frequencies of these varies in populations from different parts of Australia (Fig. 4.1). Similarly, the two antigenically distinct phenotypes, R^+ and R^-, on the surface of the red cells vary regionally, as do the two phenotypes involved in the balance of Na^+ and K^+ ions inside the red blood cells. In low potassium possums the proportion of K^+ to Na^+ in the red cells is about $13:120$ mEq./dm^3, while in high potassium possums it is $117:37$ mEq./dm^3,[19] as in most other mammals. The difference is related to the functioning of the sodium pump, which normally removes Na^+ from the red cells by active transport, involving the release of energy from ATP. Red cells from low potassium possums have the basal ATPase enzyme system but lack a second system,[17] present in high potassium possum red cells, which is activated by the presence of additional Na^+ and K^+ in the medium.

Broadly, the possums from south Western Australia and the small sample from northern Australia, near Darwin, show a high frequency for the transferrin D and R^+ phenotype (Fig. 4.1), whereas the populations east of the Nullarbor plain and central desert show a preponderance of the transferrin A allele, and the C allele is restricted to the east. This might be thought to indicate an earlier isolation of the eastern and western possum populations than of the northern and south western populations, except that the low potassium phenotype has not been found in populations from northern Australia but occurs in varying frequencies in populations in southern Australia, from 60% near Sydney, through 4% around Adelaide, to 80% near Perth. Speculations of this sort rely on an assumption that there is little selective advantage in any particular phenotype, so that the degree of difference is due to random genetic drift within isolated populations, and the longer the separation the greater the drift. However, this is not necessarily the correct explanation. The different characters may have selective advantage in particular regions and come to predominate rapidly for this reason; increase in body size, for instance, is generally supposed to be directly correlated with cooler climate as it reduces the surface area of the body relative to its volume, and in the possums of Tasmania the incidence of dark pelage has been correlated with areas of high rainfall. Then again, several other properties of the red cells differ significantly between low potassium and high potassium possums.[19] Those of the former are larger and carry more haemoglobin than high potassium cells, and are less fragile; these differences, too, may have adaptive significance in the cooler regions of southern Australia but not in northern Australia.

If the genetic diversity of possums, which these studies disclose, does reflect adaptive responses of the several populations, it may account in part for the success of this species in occupying a wide range of habitats and in adapting to European settlement and to the alien en ment of New Zealand, since it is distinguished from the other ph gerids in these three respects. It is therefore fortunate that this singu...rly successful species should also be the one about which most is known, so that many direct comparisons can be made with the macropods already discussed.

Digestion

Possums are predominantly herbivorous, but seem able to thrive on whatever vegetation is available. In eucalypt forests they eat eucalypt leaves, in New Zealand forests they eat the broad leaved tree species and the myrtaceous species of *Metrosideros*; they eat the shoots and male cones of pine trees in plantations and will come out of the scrub to graze on pastures of clover or grass; and in suburban gardens they will eat ornamental and orchard species.

Unlike in the Macropodidae, the possum's stomach is simple and relatively small, about 3% of body weight, and does not have an enlarged fore stomach free of gastric glands, so that a pH of 3–4 prevails throughout. The intestine is long and there is a large caecum which has a pH of 8·5–9·0, and which, by analogy with other non-ruminant herbivores such as the rabbit, may therefore be a fermentative chamber. The same anatomy is seen in the other phalangerids and it is therefore of interest that a form of reingestion has been observed in the koala, *Phascolarctos cinereus*, and the ringtail possum, *Pseudocheirus peregrinus*. The young koala feeds on semi-digested material from the mother's rectum at weaning[181] and it has been suggested that this may provide it with caecal bacteria; alternatively, it might be a means of providing vitamin B_{12}, as in the rabbit and rat, but if this were so it should be observed in adults as well as juveniles, which it is not, the habit ceasing when the young is fully weaned. Soft faeces, quite distinct from the normal excreta, have been observed by Dr. Griffiths to be reingested by adult ringtail possums. At some times of the day the stomach contained fresh leaves, while at other times it contained soft faecal material having a pH of 4·5. It had a high amylase activity and contained protein of bacterial origin.

The use of marker substances shows that the rate of passage of food through the possum is more variable than in the kangaroo (Fig. 3.9); this is partly because faeces are only passed at night when the possum is active, and partly because some material is retained in the caecum.[99] If plastic pellets were used, half had been passed in 16 hours, whereas 1–4 days elapsed before half the distinctive cuticle of eucalypt leaves or a soluble dye had been passed. A further complication when interpreting these results is that reingestion, if it took place, could also have extended the time of clearance of the marker. On balance, the rate of passage through the possum seems to be comparable to the quokka, although final clearance takes longer. Insufficient work has been done to determine whether the rate varies in response to dietary restriction as in sheep and kangaroos, but this comparatively slow rate of passage and the likely involvement of the caecum in fermentative digestion may account for the very high rate of digestion of crude fibre (80%) reported in a series of 3 possums fed on banana and carrot.[129] Nevertheless, other evidence suggests that the possum obtains most of its carbohydrate requirements by hydrolysis of starch rather than by fermentative breakdown of cellulose.

Carbohydrate metabolism

Amylase (or diastase) concentrations in possum blood and urine were measured 30 years ago by Anderson[14] who found comparable concentrations in both fluids (60 units/cm^3) and thus daily excretion via urine of nearly 2000 units. She did not observe any differences due to age or diet, but she did find that the rate was more than doubled in lactating possums, in which the blood sugar concentrations were also high. She measured amylase activity in various organs in order to determine the origin of the enzyme and found that, although the highest activity was in the pancreas, ligation of the pancreatic duct, which caused involution of the gland, had no effect on urinary or plasma amylase concentrations; ligation of the hepatic artery, however, led to liver degeneration and a pronounced fall in plasma and urinary amylase activity, from which she concluded that the liver is the main site of enzyme synthesis. This has recently been re-examined by Hope and Finnegan,[130] using gel electro-phoresis to separate the amylase iso-enzymes in serum, saliva and the pancreas of *Trichosurus*. They find that the species is polymorphic for serum amylase, as it is for other serum components, and that the serum iso-enzymes can be derived from a combination of the salivary and the pancreatic enzymes. They conclude that the pancreas is the main site of amylase synthesis, although some of the serum amylase may come from other sites. Its passage from the plasma to the gut is probably via saliva, as it is in the kangaroo, where parotid gland saliva has a high amylase activity, but it may also pass across the intestinal mucosa or through the pancreatic and bile ducts. Its pH optimum is 7–8 so it must be inactive in the stomach but active in the duodenum where the pH is 7–7·5.

Further breakdown of disaccharides occurs in the small intestine where the concentrations of 6 disaccharidases have been measured in tissue homogenates of a possum, a koala, and two grey kangaroos.[143] In particular, the activities of maltase, isomaltase and sucrase were 100–200 times as high in the two phalangerids as in the kangaroos (Fig. 4.2), which supports the notion that the phalangerids are dependent on carbo-hydrate absorption rather than on the volatile fatty acids of macropods.

	Maltase	Iso-maltase	Sucrase	Lactase
Trichosurus vulpecula (Adult)	41·2	22·9	6·8	0·47
Phascolarctos cinereus (Adult)	12·0	5·2	1·9	0·68
Macropus giganteus				
(1, Adult)	0·20	0·12	0	0
(2, Adult)	0·4	—	0	0·1
(3, Pouch young)	0·59	0·12	0	5·6

Fig. 4.2 Intestinal enzymic activity in units/g wet weight of mucosa, of 4 intestinal disaccharidases in the brush possum, *Trichosurus vulpecula*, the koala, *Phascolarctos cinereus*, and the eastern grey kangaroo, *Macropus giganteus*. (From Kerry[143])

The blood sugar patterns in the possum support this too (Fig. 3.7). The average values for adult male and female possums obtained by Anderson[13] were 120 mg/100 cm^3 plasma and 84 mg/100 cm^3 of packed red cells, which are in the range for non-ruminant eutherian mammals and considerably higher than Barker's[18] values of 78 mg/100 cm^3 and 20 mg/100 cm^3 respectively for plasma and red cells of the quokka. Similarly, injected glucose takes three times as long to clear from cir-culation as it does in the quokka.[126]

Adams and Bolliger[10] examined the responses of possums to reduced insulin, by destroying the Islets of Langerhans with alloxan, and the responses of intact possums to injections of insulin. Initial responses to alloxan varied from a severe fall in blood glucose to the expected rise; possums that became chronically diabetic were able to survive with blood sugar in excess of 200 mg/100 cm^3 but went into coma at levels of 400 mg/100 cm^3. Conversely, possums injected with insulin went into hypoglycaemic shock if the blood sugar level fell below 40 mg/100 cm^3. In both these respects the possum shows an opposite response to that of macropods and sheep and more nearly resembles the non-ruminant rabbit (Fig. 3.7).

Lipids

The metabolism of dietary lipids in the possum also contrasts with that of the Macropodidae.[45] The depot fats in non-ruminant mammals tend to reflect the fatty acid composition of the dietary fats, but this does not hold for ruminants such as cattle and sheep because the dietary unsaturated fatty acids, especially linolenic and linoleic acids, are hydrogenated in the rumen to form stearic acid and *trans* isomers of oleic and linoleic acids. The depot fats of the quokka, *Setonix brachyurus*, and tammar, *Macropus eugenii*, like that of ruminants, have considerable amounts of these *trans* acids in their depot fat. The fat of possums and koalas does not, but contains instead a large proportion of unsaturated C_{18} fatty acids. However, the depot fats in the possum and koala did not reflect as closely as non-ruminant eutherians the dietary levels of fatty acids, so that selective absorption must take place.

To conclude this section, the evidence from stomach anatomy, sugar metabolism and fatty acid absorption indicate that digestion in the possum differs from the Macropodidae and more closely resembles non-ruminant herbivores, such as the rabbit. If the large caecum is a fermentative chamber its role must approximate to that of the rabbit rather than to the fore stomach of the macropod; it may facilitate digestion of fibre by cellulytic bacteria and it may provide the B_{12} requirements of the animal, but at present there is no evidence either way. If this were to be pursued, the wombats would be worth comparing with the possum, since they have a simple stomach like the possum, but live almost exclusively on a diet of grass. Their teeth are highly adapted to this diet, being open-rooted and continuously growing, like those of the rabbit. There is some evidence[6] that the plains wombat, *Lasiorhinus latifrons*, may digest its food by means of cellulytic bacteria and that it is also able to recycle urea.

Thermoregulation

Trichosurus vulpecula can maintain a deep body temperature of 36–37°C against a range of ambient temperatures from 10°C to 30°C but above this temperature and up to 44°C the body temperature restabilizes at 39°C. It can continue to maintain this temperature for some hours, provided the humidity is low and water is available to drink, by panting in the same way as the kangaroos mentioned earlier. At 30°C it salivates and licks its paws and body fur but at 40°C this stops as it lies still and pants. Respiratory rate increases four fold between 30°C and 40°C and evaporative water loss increases correspondingly. Dawson[70] has analysed the relative contribution of licking, sweating and panting in the possum's thermoregulation (Fig. 4.3); if he prevented the possums from licking, their temperature was unaffected and he concluded that, since water loss from the skin increased only slightly, whereas water loss through panting increased greatly and comprised 90% of the total water loss, this latter is the main means of cooling.

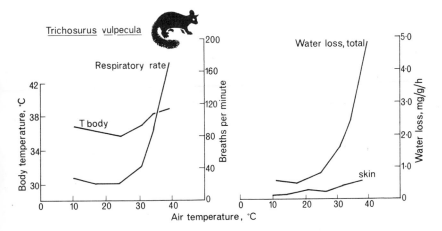

Fig. 4.3 To show how body temperature in the brush possum is maintained above 30°C by evaporative cooling, mainly from an increased respiratory rate, and only to a minor extent from the skin. (From Dawson[70])

Robinson and Morrison[221] found that the koala, and the mountain possum, *Trichosurus caninus*, responded in much the same way as the brush possum and would drink water under hot (40°C) dry conditions, but both species were more discomforted when the humidity was raised. Despite a rise in their respiratory rates, there was no change in evaporative water loss, and the ensuing stress was registered by a rise in pulse rate of both species. These responses were in sharp contrast to the cuscus, *Phalanger maculatus*, which showed no change in pulse at 40°C and high humidity, but only a 2-fold increase in respiratory rate; the evaporative water loss rose to the same rate as that of *Trichosurus caninus*, probably by sweating. However, the cuscus was quite unable to withstand dry heat and had to be removed from the experiment before the full seven hour run. This is probably related to the fact that it is an inhabitant of humid rain forest, whereas the other two species live in drier habitats.

The standard metabolism of fasted *Trichosurus vulpecula* is 12·51 kJ/h or 180·0 kJ/kg$^{0·75}$/day (Fig. 1.3) and is similar to the value for a range of marsupials and about 70% of the value for Eutheria. The lower metabolic rate of the possum correlates with blood parameters and kidney function. Blood volumes for a variety of eutherian species range between 70 and 80 cm^3/kg body weight and the plasma constitutes 41 to 54 cm^3/kg of this, so that the packed red cell volume is between 32% and 42%. This value is known as the haematocrit and is a useful parameter of changes in the body; it can vary as a result of dilution or concentration of plasma through kidney function or as a result of release of red cells from the splenic reservoir. Adrenaline from the adrenal medulla stimulates contraction of the spleen and a rise in haematocrit, whereas central nervous depressants, such as the drug chlorpromazine, cause splenic relaxation and a falling haematocrit. If total blood volume and haematocrit are measured the particular change can be determined.

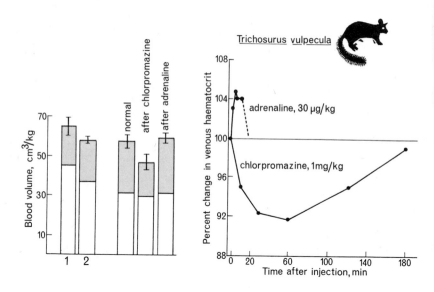

Fig. 4.4 Blood volume and haematocrit (shaded) and the effect of chlorpromazine and adrenaline on these parameters in the brush possum (After Dawson and Denny[72]). Blood volume and haematocrit in two other marsupials shown for comparison: (1) *Setonix brachyurus*[7] (2) *Didelphis marsupialis*.[56]

Three marsupials examined have smaller blood volumes than the Eutheria referred to but have haematocrit values within the same range (Fig. 4.4). Dawson and Denny[72] have examined the splenic influence on blood volume in the possum and found that its normal volume is 57·4 cm^3/kg and that the haematocrit comprises 44% of this. Injections of adrenaline induced a rapid rise in blood volume and haematocrit, whereas chloropromazine had an opposite but longer lasting effect (Fig. 4.4). In both cases the plasma volume did not change and the total volume changes recorded were due to changes in the red cell volume alone. By subtraction of the minimum from the maximum erythrocyte volumes the spleen can be seen to have changed the total blood volume by 12·0 cm^3/kg or nearly 21%, which indicates its importance as a vascular reservoir. Notwithstanding this, possums have survived splenectomy for two months without apparent ill effect.[14]

The mean blood pressure at the femoral artery of the possum is 110–130 mm Hg. This is the same as a comparable sized eutherian, the rabbit, which is surprising in view of the low values for blood volume and pulse rate observed. Since volume and pulse are both low, however, the flow rate is probably about the same as in comparable eutherians so that the relatively high blood pressure must be achieved by a more profound haemodynamic partitioning of blood between different organs of the body. We have noted the role of the spleen in this, but it could apply to many other capillary beds in the body. The total effect of many depleted capillary beds would be a lower overall metabolism, as already noted.

One anatomical peculiarity of marsupials achieves relevance in this context; it has been known for many years[7] that the brain of the possum and many other marsupials is not vascularized by the usual capillary reticulum but that the arterioles and venules are connected to each other by simple loops without anastomoses, each pair being bound within a common connective tissue sheath. Similar vascular loops have been observed in the heart and epiphyses of the euro, *Macropus robustus*, but in all other tissues a normal capillary bed was found. After damage to brain tissue, this unusual arrangement of blood vessels regenerates, which suggests that it has some functional significance in this organ. Two functions have been suggested; first, that the intimate arrangement of arteriole and venules may provide a counter current exchange effect such as we saw in the marsupial rete mirabile (Fig. 2.3) or second, and more likely in the present context, the non-capillary loop might avoid the sharp pressure drop that normally occurs across capillary anastomoses. The loops may ensure that blood flow to the brain and heart remain high despite haemodynamic shifts between other parts of the body.

Kidney function

Kidney function has been examined more thoroughly in the possum than in any other marsupial,[209] using the elegant technique of stop-flow urine analysis[169] to determine the filtration rate and the contribution of secretion and resorption in the different segments of the nephron. Anatomically the possum kidney is the same as most eutherians but it lacks the pronounced papilla of desert eutherian species and the desert-living marsupials. There are two types of nephron; the cortical nephrons have short loops of Henle situated within the cortex, while other nephrons have very long loops turning in the medulla, some of which may extend almost to the papilla. The clearance rates from the plasma to urine of solutes such as inulin were slightly lower than for the rabbit, whereas the tubular functions such as the re-absorption of glucose, sodium and water and the secretion of creatinine were similar to values obtained by stop-flow analysis on the dog and man. It would be very interesting now to subject the tammar or euro to stop-flow urine analysis, since the papilla is much longer than in the possum and the loops of Henle correspondingly longer.

To determine the maximum concentrating ability of the kidney, possums were held for 3 days without food and water. Urinary output fell from 170 cm^3/day to 23 cm^3/day and the concentration rose from 184 m Osm/dm^3 to a mean of 1061 m Osm/dm^3, the highest being 1504 m Osm/dm^3. The plasma concentration changed very little, from 303 to 316 m Osm/dm^3, so that the urine:plasma concentration ratio (U/P) is 4·8. This ratio is lower than any of the macropods measured (Fig. 3.18) and about the same as *Didelphis marsupialis* (see p. 200). Furthermore, the urinary concentration of the normal possum is about half that of the well fed quokka, which lends further weight to the idea that the macropod kidney has more effective tubular capacities than the possum and would merit the same close study. In this context it is worth noting that, although possums occur in south Western Australia, they are not found on any of the offshore islands, either those that macropods inhabit or the others (Fig. 3.15), and only occur on the much larger Kangaroo Island.

Adrenal cortex

Anderson[13] removed the adrenal glands from three male possums in a two stage operation. None of the animals showed any ill effects during the two months after the right gland was removed but all died less than two days after removal of the second.

Recently Reid and McDonald[210] extended these observations in the possum. If only one gland was removed the other gland showed a compensatory hypertrophy after 4 days, and animals were unable to survive loss of the second. The classic effects of adrenocortical insufficiency were seen; a fall in plasma Na^+, Cl^- and HCO^-_3 and a rise in plasma K^+ and plasma urea. Blood sugar remained at the normal range if food was eaten, but it fell if food was withheld or not eaten. These adverse effects of adrenalectomy could be temporarily reversed so long as injections of cortisol, desoxycorticosterone acetate (DOCA) or aldosterone were given. As well, the possums would ameliorate their Na^+ depletion by selecting 1% saline to drink.

In all these respects the possum shows the same utter dependence upon its adrenal glands for electrolyte balance and carbohydrate metabolism as eutherian species. This would not be remarkable, were it not that the American opossum shows no such dependence and can survive many months after total adrenalectomy, provided sufficient food and water are supplied to it (see p. 202). The number of observations on the opossum can leave no doubt of this conclusion but the responses of *Trichosurus* suggest that *Didelphis* is unusual in this and not that marsupials generally can dispense with their adrenals. This is supported by the only other study of adrenalectomy in a marsupial; the quokka, *Setonix brachyurus*, succumbs within two days to loss of both adrenals[57] and shows the same profound alteration in plasma electrolyte balance and blood sugar levels as the possum does.

The recent methods of identifying and measuring steroid hormones in the circulation have been applied to several marsupials. In the possum,[287] cortisol is the most abundant corticoid in the adrenal venous blood whereas corticosterone, the other main glucocorticoid, is not. A number of other intermediate products of the steroid biosynethtic pathways have been detected, including 18-hydroxycorticosterone, the precursor of aldosterone, but aldosterone has not been detected in possum blood. However, its positive effect in adrenalectomized possums would suggest that it may be there.

Cortisol is also the main steroid in *Didelphis marsupialis*,[137] in the wombat, *Vombatus ursinus*, the grey kangaroo, *Macropus giganteus*[66] and the red kangaroo, *Megaleia rufa*.[288] It is particularly interesting that it is more than four times as concentrated in the kangaroos than it is in the other three, for this may be related to the greater dependence of the macropod on gluconeogenesis (see p. 112), a process influenced by cortisol.

One unresolved facet of possum adrenal physiology was first recognized by Bourne.[47] The adrenals of female possums, as of female wombats and kangaroos, are nearly twice the size of the adrenals of males on a mg/kg body weight basis. Furthermore, the adrenals of possums enlarge during lactation and this is due to the increase in a distinct zone of the adrenal cortex which Bourne called the X zone because of a resemblance to that zone in the mouse. Chester Jones et al.,[138] analysing the adrenal venous plasma of female possums, detected cortisol at 144 µg/100 cm³ and another steroid resembling testosterone at 223 µg/100 cm³. They attributed this steroid to the X zone, since the amount in different females varied concomitantly with changes in the X zone, which in turn varied with the stage of reproduction. The function of this remains obscure; Anderson[13] observed an elevated blood sugar of 240 mg/100 cm³ in 3 lactating females, which also had enlarged adrenals and, similarly, Britton and Silvette[50] observed in lactating opossums elevated blood sugar levels that were reduced to normal levels after adrenalectomy.

The absence of aldosterone from the possums examined by Reid and McDonald[210] may have been due to the fact that they were caught near Melbourne, where the sodium concentration of the plants they ate would have been high. In another study of wombats and grey kangaroos living on the Victorian sea coast and in the Snowy Mountains[39] it was found that the mountain vegetation was deficient in salt, whereas there was an abundance in the coastal vegetation. These differences were reflected in the urinary Na^+ concentration, the thickness of the zona glomerulosa of the adrenal cortex and the concentrations of aldosterone in the peripheral circulation of the several species. Aldosterone concentration in blood was 4 times as high in the alpine kangaroos (40 ng/100 cm³) as in the coastal kangaroos (9 ng/100 cm³) and its concentration in alpine wombats was double that of coastal wombats. From these results it would seem that the search for aldosterone in the possum should be directed to alpine or salt depleted possums rather than to coastal animals.

ECOLOGY OF THREE PHALANGERIDS

Among the macropods that were considered in Chapter 3, the extreme conditions of the environment imposed limitations upon the populations, either by a cessation of breeding or by a more or less severe mortality of pouch young and independent juveniles. Social interaction between members of the same population does not appear to be important, at least in the quokka or red kangaroo, although it might be in the euro.

Most phalangerids live in more benign environments where regular winter rainfall and the rich eucalypt forests provide a habitat with much less extreme fluctuations in the conditions for life. Populations of these species are unlikely to be controlled by adverse climatic conditions, so other factors must operate to maintain the fairly stable populations encountered. Three species can be considered with this question in mind. Populations of the brush possum, *Trichosurus vulpecula*, have been studied in Victoria, Canberra and near Brisbane, as well as in New Zealand; the ringtail possum, *Pseudocheirus peregrinus*, has been studied in Victoria; and the greater glider, *Schoinobates volans*, has been studied near Canberra and in Queensland. A summary of these several studies is given in Fig. 4 5.

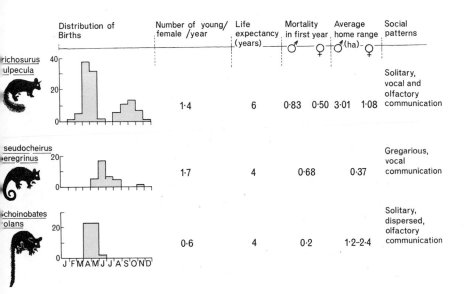

Fig. 4.5 A comparison of breeding patterns, fecundity, longevity and social behaviour in three species of the Phalangeridae. (From Dunnet,[80] Hughes, Thomson and Owen,[136] Smith,[261] and Tyndale-Biscoe and Smith[281])

The brush possum, Trichosurus vulpecula

In the brush possum the majority of births occur in April and May and the single young is suckled in the pouch for five months. If it dies prematurely it will usually be replaced, so that about 90% of the females produce at least one young per year. When the young leaves the pouch it rides on its mother's back for another 2 months but during this time the mother may sometimes return to oestrus, so that a small second peak of births occurs in September and October. The contribution of the second peak varies according to place and time: there is none in Tasmania, where breeding starts one month later, it is seasonally variable in New Zealand,[98] while near Canberra and in New South Wales about half the females produce a second offspring at this time. Thus, in the Canberra population studied by Dunnet[80] the average number of young produced per female each year was 1.4. At this rate the adult population could potentially be replaced if each female only bred for two years. Since females reach sexual maturity at one year, it might be expected from

Plate 5   Brush possum, *Trichosurus vulpecula,* from southern New South Wales. (Ivan Fox)

this that the population turnover would be rapid and life expectancy about 3 years. This was not found to be so and most of the adults remained throughout the two year study and three were recaught five years later in the same localities.[166] I also observed this in New Zealand possums. A sample of possums can be sorted into three age classes by the degree of fusion of the epiphysis of the tibia[144] and in samples of possums shot or trapped on Banks Peninsula, New Zealand, the relative proportions of possums less than 1 year old, 1–4 years and more than 4 years was determined (Fig. 4.6). In one area (b), where there had been a previous history of severe trapping, the young animals composed one third of the population whereas in another undisturbed area (a), they composed only 8%, and those over 4 years composed 55%. In the latter population such a proportion could only result if adult animals lived on average for more than 6 years. In a stable population of possums, therefore, the realized fertility must be severely reduced by mortality of young animals but if mortality is imposed on the population by trapping, as in the other area, juvenile mortality will have been relaxed and more of the reproductive potential realized.

In Dunnet's [80] study all the animals on the study area were marked and frequently recaptured in box traps. From this information he concluded that some of the population consisted of resident animals, which were frequently recaptured within well defined ranges, while others, which might be caught very infrequently, he called transients. Transients were

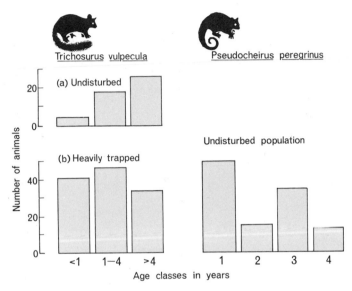

Fig. 4.6 Age composition of two separate populations of brush possums and one undisturbed population of ringtail possums. (Own data for brush possum, Thomson and Owen[271] for ringtail possum)

invariably young animals less than a year old and usually immature, whereas residents were breeding animals more than a year old. No young males were residents but a few young females became residents. Resident females occupied home ranges of about 1·08 hectares but did not exclude other animals from the area, so it could not be termed a territory. On the other hand, the resident males did defend their territories against other males, especially young males. These territories were about 3·01 hectares and completely overlapped the smaller home ranges of females. Since females were not exclusive about their range, the density in an area could be quite high, whereas male territories being exclusive the density over the whole study area was fairly uniform. As a consequence of this, males composed only about 40% of the population throughout the study area.

J. W. Winter has studied the behaviour of free-living, tagged, possums by direct observation at night and has allowed me to refer to his work. During one year the 500 hours of observation were distributed throughout all seasons, while in the second year the same amount of time was concentrated at important periods, such as the autumn breeding season and the time of separation of juveniles from their mothers.

He found the same overlap of home ranges that Dunnet had found but did not find that males would defend their areas against other males. His study confirmed that adult possums are quite solitary and sleep in separate dens during the daytime. At night-time, encounters between adults of either sex are mild, the usual response being for both animals to avoid direct contact, but to peer silently at each other with erect ears. Avoidance is probably also helped by possums informing one another of their presence by marking branches and other objects with scent (Fig.

(a) *Vocal*

Call	Caller	Situation	Probable message	Response from others
Short chatter	♂ ♀		alarm	alert or chatter
Screech	♂ ♀ juvenile	encounter	threat	flight or attack
Zook-zook	juvenile	separated from mother	lost	approach of mother
Click	♂	following oestrous ♀	appease-ment	reduced agonistic response of ♀
Shook-shook				
Long chatter	♂	post-coital	—	none apparent

(b) *Olfactory*

Gland	done by	behaviour	possible message	response from others
Sebaceous, chin	♂	chinning	specific location	no response apparent
Sebaceous, chest	♂ ♀	chesting		
Holocrine, cloacal	♂ ♀	urine dribbling	general location	
Apocrine, cloacal	♂ ♀	ejection	fear	

Fig. 4.7 Vocal communication in *Trichosurus vulpecula*, and olfactory behaviour that may be used for communication. (Personal communication, J. W. Winter, University of Queensland, by permission)

4.7). Direct evidence of the importance of scent has not been established, but possums have quite a repertoire of scent organs, and these are best developed in males (see p. 39). There are modified sebaceous glands on the chin and the chest, as well as two sets of parcloacal glands. One of these produces a holocrine secretion which is distributed in urine, and urine dribbling is a common form of marking. The other, an apocrine secretion, seems to be ejected only as a reaction to extreme fear. The only area that a possum actively defends is its den, and it responds to an intruder with a loud screech.

There is no pattern of mutual grooming or other contact activities such as occur in gregarious mammals and the only long term contacts are between a female and her offspring. After the young leaves the pouch it clings to the mother's back for another two months. If it becomes separated during this time it makes a characteristic call to which the mother responds by returning to it and by discouraging other possums from approaching it. Later, close contact gradually declines, and by nine months the mother shows increasing antagonism towards the juvenile, which is eventually driven from the den.

Agonistic behaviour between adult males and females is only reduced at oestrus. Males are attracted to an oestrous female, but the female does not solicit the male and remains averse to contact. The male

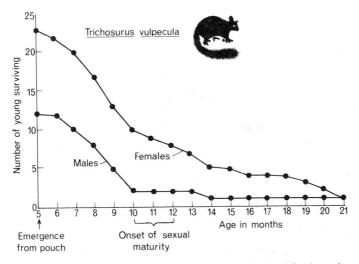

Fig. 4.8 The numbers of native young of each sex that survived on the study area to each successive month after emergence from the pouch at 5 months. Total of males below the lower line and of females between the two lines. (From Dunnet[80])

uses a call at this time that closely resembles the 'lost' call of a juvenile, and it may be the effective signal in reducing agonistic responses of the female sufficiently to allow copulation to occur.

Thus several attributes keep adult possums separated from one another and keep the population dispersed. When juveniles are driven from their mothers' dens they are subordinate to the resident adults and so they tend to move out of their natal area. Dunnet's[80] transients were predominantly of this age group. If the habitat is fully occupied the result must be that most of the juveniles die or occupy less favourable country elsewhere. This is clearly shown in the disappearance of young animals from Dunnet's study area (Fig. 4.8), in which the major decline occurs at 9–10 months of age and falls more heavily on males than on females. Nevertheless, the numbers of both sexes declined so that only 2·6% of the males and 8% of the females eventually became established as residents.

The brush possum has a breeding potential well in excess of requirements for replacement of the population, and the size of the population is regulated by behavioural factors that cause dispersal of the adults and heavy mortality of the newly independent young. When the habitat is not fully occupied, as occurred in New Zealand after the first liberation of possums there, this control is lifted with dramatic results in the growth of the population. As an example of this,[291] 36 possums were liberated in the Catlins district of South Island in 1894–5; twelve years later 60 000 possum skins were taken from that district by trappers.

The ringtail possum, *Pseudocheirus peregrinus*

The ringtail possum is also arboreal, nocturnal and herbivorous but it is more restricted to forest habitat than is *Trichosurus vulpecula*; it, too, was introduced to New Zealand but failed to become established there.

Unlike brush possums, ringtails are gregarious and their social behaviour is centred upon the communal dreys, which they build in the branches of small trees. In a study of ringtails in Victoria, Thomson and Owen[271] found that the communal group usually consisted of an adult male with one or two adult females and their dependent progeny and the immature offspring of the previous year. Such a group might build several nests at different heights and localities and would repulse any strange ringtail that attempted to enter an occupied drey. The distribution and abundance of dreys was highly correlated with particular types of habitat; they were most abundant in low scrub, or areas regenerating after partial clearance, but were much less common in heavily timbered country where the understory was sparse. The requirements appear to be the congruence of suitable sites for dreys and preferred food plants.

Plate 6 Female ringtail possum, *Pseudocheirus peregrinus*, with 4 month old juvenile on her back, from southern New South Wales in October. (Ivan Fox)

Since dreys are essential for the survival of young after they relinquish the mother's back, as well as being a daytime refuge for adults, requirements for drey sites correlate closely with abundance of ringtails.

The breeding potential of the ringtail is higher than that of the brush possum. Females attain sexual maturity at one year and, like in the brush possum, most births occur in May-June with a very minor second peak in November (Fig. 4.5). Similarly, young that are lost prematurely are replaced, so that about 90% of the females successfully produce young each season. The average litter size is 1·9 and the mode is 2, so that the average number of offspring per female per year is 1·7, which is higher than that of the brush possum. However, the life expectancy is considerably shorter than the brush possum's, and few ringtails survive their fourth year (Fig. 4.6), so that population turnover is more rapid. This comes about by a fairly severe mortality (68%) in the first year, followed by a lower mortality in the second and third year but a sharp increase in the spring and summer of the fourth year (Fig. 4.9), when the environment is least favourable to the animals. The first year mortality is not more severe on males than females, so that the sex ratio remains at parity throughout the population.

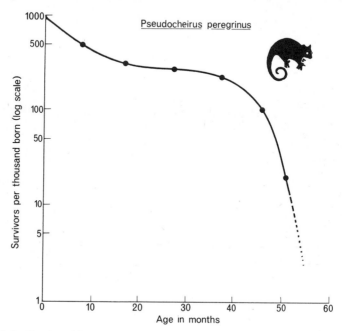

Fig. 4.9 Survivorship curve calculated for a Victorian population of ringtail possums. (From Thomson and Owen[271])

The greater glider, Schoinobates volans

The third phalangerid to be considered shows yet another pattern. *Schoinobates volans* is closely related to the ringtail (Fig. 1.9) but its adaptations for gliding enable it to live in the largest forest trees. Its food preferences are similar to the ringtail's but it is only rarely found in small trees or regenerating saplings; it feeds exclusively on the canopy foliage at night and lives in hollow spouts by day. It is most abundant in wet sclerophyll forest where ringtails are uncommon and is rare in dry sclerophyll forests, which ringtails prefer.

In our studies on this species near Canberra[281] we have found that gliders, like possums, are solitary for most of the year. A nearest neighbour analysis of their daytime distribution, obtained when forest was being felled, showed that the total distribution, and that of each sex separately, was more uniform than a random distribution and was

Plate 7 Female greater glider, *Schoinobates volans*, with young in the pouch. The light grey pelage of this animal is the less common colour phase (about 20%) in the forests of southern New South Wales. (Stan and Kay Breeden)

not clustered as that of a gregarious species would be. This suggests that there may be active repulsion between members of the population and more especially between members of the same sex. Unlike possums, gliders are generally silent, so that if this spacing has a behavioural basis it is likely to be olfactory, since the paracloacal glands are well developed, especially in males. Direct observations on behaviour have not been made on this species so the means by which the spacing comes about and the home range of individual animals remains conjectural. The overall density, both near Canberra and at the northern end of its range in Queensland, is about one adult per 1·2–2·4 hectare (Fig. 4.5), which is similar to the possums in savannah woodland but considerably higher than possums in the same forest as the gliders.

Like the other two species, *Schoinobates* is polyoestrous and has a restricted season of birth in April-June (Fig. 4.5), but unlike in them, young lost prematurely are not replaced and there is no second peak of breeding after the first young is weaned in December. In consequence the fecundity is much lower, only 60–75% of the adult females producing a single offspring, and the reproductive rate for the population of 0·67/female/year is half that of the possum and almost one third that of the ringtail (Fig. 4.5). The females that fail to rear an offspring may have lost their pouch young prematurely but there is also evidence that some adult females each year do not breed at all. Although a few females may breed in the season following birth, most do not breed until the second season after birth, so on these two counts the population could not be maintained unless females lived for an average of four years. Since we have no way at present of determining the age of gliders, we do not know how long they live.

The mortality during the first year is much less than in the other two species, being only about 20%, but it is distributed very unevenly between the sexes. The sex ratio at birth and among young in the pouch is 50% male but this changes at the time of weaning to about 40% male. This ratio prevails in all older classes and in all samples collected near Canberra and in one from Queensland. It thus appears to be a common feature in the population structure of the species. It is also remarkable that in all populations sampled the number of adult males is about equal to the number of breeding females. We have been led to postulate from this that the male specific juvenile mortality is density dependent, being effected by a behavioural factor similar to the one in the possum, and that the number of adult males directly determines, on a 1:1 basis, the number of females that will breed in any year. We have some evidence to support the first but little so far to support the second. The hypothesis suggests a way in which the population size could be regulated, by controlling natality through behavioural responses to density, rather than by severe mortality of a reproductive surplus, as in the other two species.

Whatever the final fate of this hypothesis may be it is evident that *Schoinobates* has a lower fecundity and lower juvenile mortality than the other two species examined and that this pattern results in a stable population, well-spaced throughout the forest.

CONCLUSION

Each of these species of phalangerids achieves population stability by different means; either by varying the reproductive potential, or the mortality in different age classes, or by exploiting different behaviour patterns of a social or solitary life.

Other studies on other phalangerids might show other patterns. A study on the Papuan sugar glider, *Petaurus breviceps*, by Schultz-Westrum[229] has disclosed that scent plays a very important role in maintaining social cohesion in this species by identifying members of a group with the scent of the dominant male, which is secreted by glands on his forehead. This is in direct contrast to the use of scent by *Trichosurus* and also contrasts with the lack of olfactory communication in the social *Pseudocheirus*. The common wombat, *Vombatus ursinus*, like *Pseudocheirus*, has a life centred on fixed home sites, in this case the burrows it excavates. Its social pattern seems to be similar to *Trichosurus*, in that contact between adults is minimal and the longest associations are those between a female and her offspring.

5

Small Marsupials—the Burramyidae and some Dasyuridae

PHYSIOLOGICAL ADAPTATIONS OF SMALL HOMEOTHERMS

For all animals there is an important relationship between body size and metabolic rate, which is much more pronounced for homeothermic vertebrates, the birds and mammals, than other animals. In its simplest terms the weight of the body increases by the cubic power whereas the surface area increases by the square power. The strength of muscle and bone depends on the cross sectional area, so the strengths of these tissues increase by the square power also. Thus a comparison between a small and a large mammal shows the small one to have proportionately a much greater surface area and to be proportionately much stronger than the large one. The greater surface area imposes greater demands upon the body in heat transfer from or to the environment and in evaporative water loss from lungs and skin; this greater energy loss must be made good from the food intake and less of the food is available for synthesis into stored material, so small animals cannot survive starvation for as long as large ones can. For example, a mouse consumes the equivalent of about 25% of its body weight each day, whereas a man consumes about 1%, so a mouse cannot survive more than a few days without food, whereas a man can survive for several weeks.

For the same reason, as Kleiber[149] vividly showed, one large animal takes much longer to exhaust its food supplies than an equivalent biomass of many small animals.

There is thus in body size a nice balance of advantages and disadvantages; under favourable conditions the species of small size converts food more rapidly and the population proliferates faster than the large sized species, but in adverse times a small sized species will decline rapidly as its members succumb to the lack of food or the adverse environmental conditions, whereas members of the large species can withstand adversity for very much longer.

Many species of small mammal and birds (i.e. less than 2 kg body weight), however, ameliorate the disadvantages of small size by eating concentrated foods, such as nectar, seeds or insects, and by entering a state of reduced metabolism, variously known as torpor, aestivation or hibernation. This state of torpor involves a considerable reduction in activity and body functions but is distinguished from the cold torpor of lower vertebrates by the animals' ability to arouse from torpor to full thermoregulation even while the ambient temperature remains low. This state has been considered by some to be a phylogenetically primitive form of thermoregulation, but the fact that it seems to occur only in those representatives of several orders of mammals that are of small size and live in areas with climatic extremes, argues that it is a physiological adaptation to these conditions. Among Eutheria the phenomenon occurs in the Rodentia, Insectivora and Chiroptera, and among marsupials it is known to occur in the small members of three families.

Hudson and Bartholomew[132] have proposed three degrees of adaptation to torpor. In the least developed form the species shows a daily fluctuation in body temperature (T_B) and basal metabolism, but insufficient to stop feeding or excretion. At the low point in the diurnal cycle such an animal need only stop shivering in order to cool to a T_B characteristic of torpor. Such facultative torpor may occur in response to food shortage but the species cannot survive more than a few hours in torpor nor tolerate a T_B less than about 15°C.

Obligate daily torpor is the next stage and is seen in desert adapted species that avoid high daytime temperatures and consequent water loss by entering a burrow. This reduces evaporative water loss immediately and by dropping T_B to T_{Burrow} a further saving is achieved in reduced respiratory rate; this can result in a reduction of total water loss to 3·5% of that on the surface at normal T_B, a very significant saving.

The final stage is true hibernation, in which torpor lasts for extended periods, and is an adaptation to low ambient temperatures. All hibernating mammals arouse from torpor occasionally and some do regularly. It is thought that metabolic products, especially urea, accumulate during torpor but, since the kidney cannot function at the low temperature of torpor, arousal is necessary for it to discharge them. But arousal from torpor is very costly in energy, being equivalent to about 10 days of torpor. Hibernating species of more than 200 g body weight can store sufficient fat to provide their total energy requirements during hibernation, including those of periodic arousal, but species smaller than this are unable to carry sufficient reserves and they are obliged to eat and drink as well as excrete during regular periods of arousal.

How are small marsupials adapted to the limitations of size? The pigmy possums, Burramyidae, some of the smaller Dasyuridae and the didelphid, *Marmosa microtarsus*, have been examined.

Plate 8   The pigmy possum, *Cercartetus nanus*, from Queensland. (Stan and Kay Breeden)

THE BURRAMYIDAE

The Burramyidae includes the pigmy possums, *Cercartetus*, and the pigmy glider, *Acrobates pygmaeus*, which weigh less than 100 g, and the very rare and somewhat larger *Burramys parvus*. They are closely related serologically and cytologically[111] and are thought to resemble most nearly the ancestral phalangerids and to be the most closely related, within this group, to the Dasyuridae. The first two species are known to become torpid but it is not known whether *Burramys* does, although it must be exposed to low winter temperatures in its alpine habitat in the Snowy Mountains.

Cercartetus

Daily and seasonal activity

Four species of *Cercartetus* occur in the wetter parts of Australia and in Tasmania. They are solitary, nocturnal animals which live in forest trees and are seldom seen except when dislodged from the rotten centre of an old tree at felling. They eat insect larvae and adult beetles and

Plate 9 The fat-tailed dunnart, *Sminthopsis crassicaudata*. (Edric Slater)

moths, which are caught in flight, as well as nectar. In captivity their daily food intake, while active, is about 7% body weight and on this they become obese. If exposed to ambient temperatures, they feed much less in the winter months and are torpid on most of the days (Fig. 5.1). The Hickmans[118] observed two specimens of *Cercartetus nanus* and two of *Cercartetus lepidus* each day for a year and found that periods of torpor and activity alternated; in the spring and summer the periods of activity lasted several days, alternating with torpor for one or two, whereas the pattern was reversed in autumn and winter. Thus torpor in *Cercartetus* appears to be associated with low ambient temperatures and no specimen was torpid when T_A was greater than 19°C, but like hibernating eutherians, they could arouse from torpor and be active at temperatures as low as 2°C. It is unlikely that torpor is associated with a shortage of insects, rather than temperature, since it occurs in animals held captive outside that are provided with an abundance of food, including mealworm larvae. The Hickmans did not provide information on the weights of their specimens, but four specimens of *Cercartetus nanus* from New South Wales, which I kept for three years, showed regular changes in weight

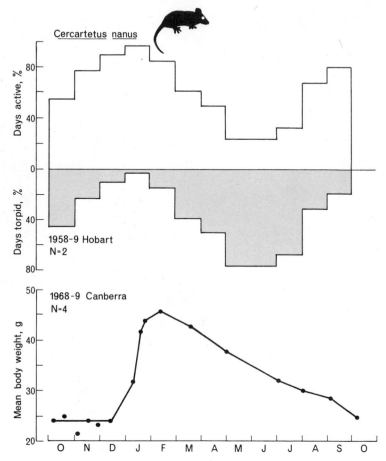

Fig. 5.1   Seasonal changes in activity and body weight of pigmy possums kept outside. (Activity data from Hickman and Hickman[118])

each year (Fig. 5.1), although food was provided all the time. The weights increased suddenly each December when the animals were active, stayed high until autumn and then steadily declined to half the summer value by September.

Thermoregulation

Thermoregulation in the active and torpid states of *Cercartetus nanus* has been investigated by Bartholomew and Hudson[29] on 9 specimens from Tasmania, which they maintained in constant conditions of photoperiod (12 h) and temperature (20–25°C). The animals became obese in

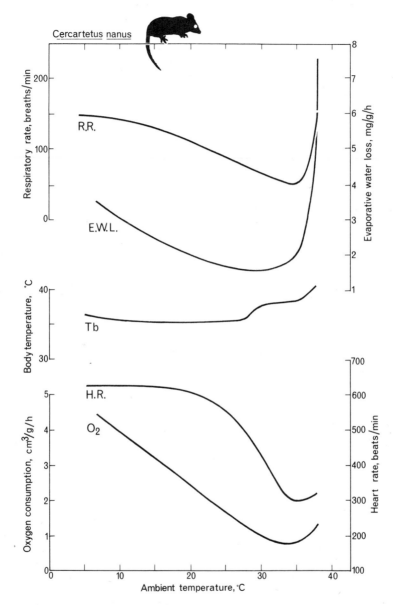

Fig. 5.2 Responses of the pigmy possum to changing ambient temperature. Note that the minimum values for all parameters except body temperature occur at ambient temperatures of 30°–35°C. Compare to the responses of the much larger brush possum in Fig. 4.3. (Redrawn from Bartholomew and Hudson[29])

these conditions and did not show the annual cycle of increased torpor in the winter months that the Hickmans' animals displayed, but they continued to enter torpor spontaneously but randomly.

Pigmy possums have a very labile body temperature, which correlates with different states of activity. In sleeping animals T_B is maintained at 27°C to 31°C while in the fully awakened state the animals maintained T_B between 32°C and 38°C at ambient temperatures from 5°C to 30°C; above this they showed signs of heat stress and T_B rose to 40°C (Fig. 5.2). The minimum oxygen consumed by quiet, wakeful animals was between T_A 31°C and 35°C and consumption rose at temperatures above and below this; this is the thermoneutral zone, where minimum energy is expended by the animal to maintain T_B and it was also reflected in the respiratory rate and the heart rate (Fig. 5.2). The respiratory rate rose steeply above 35°C as evaporative cooling began, as shown by a 5-fold increase in water loss. It is interesting to compare these responses to those of the much larger brush possum, *Trichosurus vulpecula* (Fig. 4.3). At temperatures below the thermoneutral zone of this species neither the respiratory rate nor the evaporative water loss are elevated, as they are in the pigmy possum, but both rise faster at temperatures above 25°C. Because of its larger size, heat loss is less than in *Cercartetus*, which is advantageous to it at low, but not at high, ambient temperatures.

These several measurements on *Cercartetus* thus clearly show how the maintenance of body temperature and hence activity outside the thermoneutral zone is only achieved at great cost to the animal's energy and water reserves. This cost can be avoided at T_A below 30°C by entering torpor and allowing T_B to follow T_A.

Hibernation or torpor

A pigmy possum in torpor is curled up in a tight ball with the long tail coiled on one side. The large ears are limp and folded, and the animal feels cold to the touch. The T_B is usually a few degrees above ambient and the oxygen consumption is reduced to less than 10% of the active state (Fig. 5.3) and respiration is correspondingly reduced, with periods of apnoea. The heart rate becomes very slow, the comparison being most marked at low ambient temperatures; thus at $T_A = 5$°C the heart rate was 650 beats/minute in the active state and only 60–80 beats/minute in torpor. At this low rate the electrocardiogram showed a normal pattern without irregularity or missed beats. The ability of the heart to function properly at low body temperature is one of the most important adaptations for hibernation and one which distinguishes natural hibernators from non-hibernators, such as the rat or opossum in experimental hypothermia (see p. 196).

The other important adaptation for hibernation is the ability to arouse spontaneously in the absence of any amelioration of T_A. Pigmy possums will remain torpid for several days, the longest time recorded by the Hickmans[118] being 12 days. It is not known what precipitates spontaneous arousal, but pigmy possums will come out of torpor if disturbed. At first disturbance the torpid animal utters a faint hiss, which indicates that some sensory perception is functional. During arousal there is an activation of the heat producing mechanisms with a concomitant rise in oxygen consumption (Fig. 5.3) and body temperature. The rate of rise of T_B is about 0·33°C/minute, which, again, is comparable to the rates of

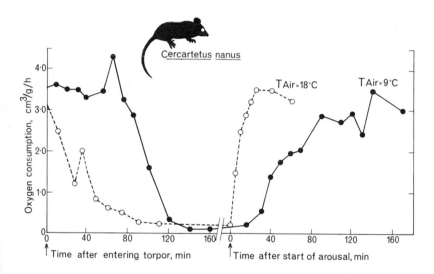

Fig. 5.3 Changes in oxygen consumption during torpor and during arousal in the pigmy possum, when the ambient temperature was held constant at 9°C (●—●) and at 18°C (○- -○). Body temperature at the start of arousal was 12°C and 20·1°C respectively and reached 33°C in both cases at peak oxygen consumption. (After Bartholomew and Hudson[29])

eutherian hibernators during arousal. In one respect, however, the pigmy possum differs; in many eutherians the anterior end of the body warms up faster than the posterior due to the release of heat from the thermogenic brown fat that lies across the shoulders, but in *Cercartetus* oesophageal and rectal temperatures rise at the same rate. It is not known whether brown fat is involved in arousal of *Cercartetus*, or indeed, whether it occurs in these animals at all.

Reproduction

Reproductive activity is usually minimal during true hibernation and, in fact, gonadal inactivity has been considered by some to be a criterion of hibernation as distinct from dormancy. It might be supposed, therefore, that in *Cercartetus*, with its markedly seasonal pattern of torpor, reproduction would be seasonal too. The rather sparse records suggest that it is not, but unfortunately these all relate to *Cercartetus concinnus* from Western and South Australia and not to *Cercartetus nanus*. Of the twelve records of breeding females, 8 were collected in spring and summer and in all cases the female was carrying 3–5 young in the pouch, and in all save two collected in winter, the females were simultaneously pregnant. It is known from several independent observations that lactating pigmy possums kept isolated from adult males will give birth to a second litter when the first litter vacate the pouch at 50 days and it was supposed that this was due to supervention of diapause in the second set of embryos (Fig. 2.11). Clark's[63] study of all the available museum material disclosed that intrauterine development continues concurrently with lactation, albeit very slowly. If conception occurs post-partum, as seems likely, then the gestation period is 50 days, which is the longest gestation of any marsupial not undergoing embryonic diapause (Fig. 2.5). It is not known how long gestation is in non-lactating females, since the species has not been bred in captivity, but the limited data suggest two possibilities. If the long gestation only occurs in lactating pigmy possums, then it may not be due to a total stasis as in macropods, but to a slowing of embryonic development, such as occurs in the roe deer and the European badger. Conversely, if the long gestation prevails in non-lactating females, its unusual length may result from periodic halts during torpor. This second alternative would be very unusual, if not unique in mammals.

The closely related pigmy glider, *Acrobates pygmaeus*, as mentioned above, is also thought to undergo periods of torpor, so it is of some interest that the only lactating female of this species to be examined and described[121] had 'blastodermic vesicles in the uteri and three young in the pouch'.

It is clear from the foregoing that the pigmy possums would repay a close study, which attempted to correlate nutritional, physiological and reproductive functions in the context of the natural environment.

THE DASYURIDAE

Size and aestivation

The Dasyuridae are a fairly uniform group taxonomically, the differences between species being mainly due to size. All the members are carnivorous and range in size from *Sminthopsis* at 10–20 g, to *Sarcophilus harrisii* at nearly 10 kg; they occur in all the main habitats of Australia. When the thermoregulation of several species was compared,[187, 221] the importance of size was clearly demonstrated; all the species maintained a steady T_B when exposed to ambient temperatures from 5°C to 30°C but, since oxygen consumption was not measured, it is not possible to say what the relative energy expenditures between the different sized species was, nor where their respective thermoneutral zones are.

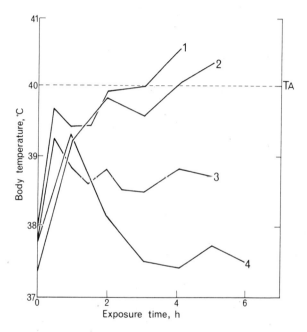

Fig. 5.4 Responses of four species of Dasyuridae to an ambient temperature of 40°C and 30 mm Hg humidity. 1. *Antechinus flavipes*; 2. *Sminthopsis sp.*; 3. *Dasycercus cristicauda*; 4. *Sarcophilus harrisii*. (From Robinson and Morrison[221])

Differences correlated with size were apparent in the several responses to ambient temperatures of 40°C and moderate humidity (Fig. 5.4). The largest species, *Sarcophilus harrisii*, maintained a normal T_B of 37·5°C for 6 hours; it showed no distress and did not pant or wet its pelage but it did drink water, indicating that evaporative cooling by sweating is probably the main route for heat dissipation in this species. The other smaller species showed varying degrees of distress, lying still and breathing fast, and T_B rose almost to 40°C. The largest, *Dasyurus hallucatus* (650 g), stabilized its T_B at 39·5°C for the 6 hours duration of the experiment, as did the desert-living *Dasycercus cristicauda* (72 g), but both drank water. The smallest species (Fig. 5.4) could not cope with the high temperature

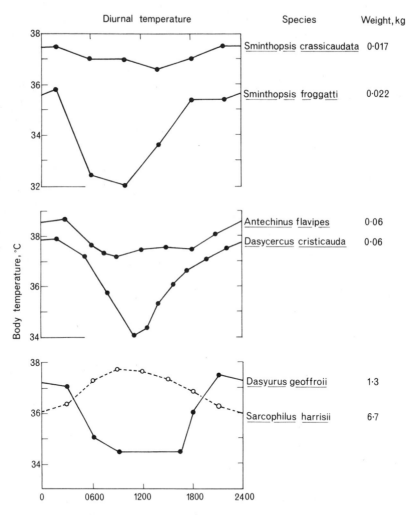

Fig. 5.5 Diurnal fluctuations in mean body temperature of 6 species of Dasyuridae. Five species are nocturnal and only *Sarcophilus harrisii* is active by day; *Sminthopsis froggatti* and *Dasycercus cristicauda* become torpid each day, the other species do not. (After Arnold and Shield,[15] Godfrey,[100] and Morrison[187])

and had to be withdrawn after 2 or 4 hours when T_B rose above 40°C.

Despite their inabilities to endure a temperature of 40°C, several of these species live in country where such temperatures occur often. Their adaptations for survival include avoiding direct exposure, a marked diurnal temperature cycle and torpor. Only *Sarcophilus* is active in the day; all the others are nocturnal and several live in burrows or other shelters during the day, where the microclimate is less extreme. As well as this, they all show a diurnal fluctuation in T_B with the lowest point at mid-day (Fig. 5.5). In some, such as *Dasycercus cristicauda* and *Sminthopsis froggatti*, the temperature range is 3–4°C, whereas in *Sminthopsis crassicaudata* and *Antechinus flavipes* it is less than one degree. These differences are interesting since the former species regularly become torpid, whereas *Sminthopsis crassicaudata* seldom does and *Antechinus flavipes* never. If we look at these four species in more detail we can see how this is related to their several life patterns.

Desert species

Dasycercus cristicauda

The mulgara, *Dasycercus cristicauda*, lives in the arid parts of Central Australia in sandy spinifex country. The annual rainfall is less than 300 mm, so free water is generally not available and the mulgara's main source of water must be its food. It eats insects and small vertebrates and, in captivity, mulgaras consumed 20–25% of their body weight in meat each day, which is normal for an animal of its size (50–100 g). Water conservation is clearly important to it, and the two main avenues of water loss are by evaporation and urinary excretion. Captive mulgaras kept at 25–30°C without access to water and fed on lean meat or whole mice, were able to maintain water balance and body weight, despite the need to excrete considerable amounts of urea.[228] They did this by excreting a highly concentrated urine; the urea concentration reached 2610 m M, or about 8 times the plasma urea concentration, and the electrolyte concentration was also high, ranging from 192 to a maximum of 674 m Eq/dm³.

These values show that the mulgara's kidney has a concentrating ability exceeding that of eutherian carnivores and the desert macropods (Fig. 3.18), and is in the same class as that of the desert-adapted rodents such as *Dipodomys merriami*. Water loss by evaporation has not been measured in mulgaras but they live by day in subterranean burrows where heat load would be avoided and evaporative loss probably reduced by a higher humidity. If they also enter torpor, as the laboratory observations indicate, this would enable even greater conservation of water, as was shown to hold for species in Hudson and Bartholomew's[132] second grade of adaptation to torpor.

Sminthopsis crassicaudata and Sminthopsis froggatti

The role of torpor in the daily cycle has been more fully examined in comparative studies between *Sminthopsis froggatti*, which lives in similar country to the mulgara, and *Sminthopsis crassicaudata*, which lives in a more vegetated habitat. *Sminthopsis froggatti* is the larger species (16–24 g) and, like the mulgara, feeds on insects and small lizards, whereas adults of *Sminthopsis crassicaudata* weigh 12–18 g and live entirely on small insects and other terrestrial arthropods; they avoid lizards or make only perfunctory attempts to catch them. The different feeding patterns of the two species seem to be related to their respective diurnal activity.

Crowcroft and Godfrey[69] measured the activity of individual animals by recording, with a sensitive microphone placed under the cage, the sounds they made while active. The skin temperature of the undisturbed animals was also recorded via a thermocouple fitted inside the small nest chamber in which each animal rested.[100]

Both species showed only a minimum of activity during the day, commensurate with periodic urination and defaecation, while the peak of activity occurred during the first two hours after dark. In *Sminthopsis crassicaudata* activity continued at almost the same intensity all night as the animals engaged in feeding, exploring and sexual encounters, whereas the *Sminthopsis froggatti* completed their feeding during the first 2–4 hours of darkness and their activity then declined to daytime levels. They would enter torpor between 0200 and 0600 hours, as determined by a fall of 8–12°C in skin temperature; if food was withheld they became torpid earlier and remained torpid longer than if food was plentiful. *Sminthopsis crassicaudata*, on the other hand, never became torpid if food was plentiful, even at ambient temperatures of 5°C, but if food was withheld for two days, activity in the second half of the night declined on the second and third night, while activity in the first half intensified, thus coming to resemble the other species' pattern. On the second night the animals would become torpid by about 0800 hours and remain so for a few hours. If starvation continued the period of torpor lengthened and some animals failed to arouse and perished.

As mentioned earlier, *Sminthopsis froggatti* has a wider diurnal fluctuation in T_B than *Sminthopsis crassicaudata*, so that it can readily enter the torpid state by allowing the normal morning decline to continue below a T_B of 28°C. For *Sminthopsis crassicaudata* a fall in T_B to the torpid state is much further below the normal diurnal low point but can occur under the stress of starvation.

Arousal from torpor in both species is about the same rate as in *Cercartetus*, namely 0·3 to 0·5°C/min, but beyond this nothing is known about the thermogenic activities in these species.

In comparing the two species, *Sminthopsis froggatti*, the larger animal, can obtain its food quickly by catching a few large insects or a lizard and can then conserve energy by entering torpor, whereas *Sminthopsis crassicaudata* must spend more time foraging for the small insects that make its diet, so that it cannot afford to become torpid. However, it can store fat in its tail to a greater extent than *Sminthopsis froggatti*; by drawing upon this it survives temporary food shortage and only enters torpor as an emergency measure.

The reproductive patterns of the two species are similar. Both are polyoestrous and breed during most of the year, probably in response to favourable conditions when these occur. The gestation is 16 days in *Sminthopsis crassicaudata* and 12·5 days in *Sminthopsis froggatti*,[101] which indicates that the regular daily torpor of the latter does not affect the rate of intrauterine development, a point of contrast with *Cercartetus* mentioned earlier. The litter size is about 8, lactation lasts about 70 days and the young are sexually mature at 6 months. Thus these species have a high reproductive potential and the patterns of continuous breeding resemble the reproductive adaptations of the desert kangaroos and contrast with the monoestrous and rigidly seasonal breeding of the dasyurids of the more temperate habitats, as exemplified by *Antechinus stuartii*.

Notoryctes typhlops

Before considering *Antechinus stuartii*, brief mention should be made of that most rare and unusual animal the marsupial mole, *Notoryctes typhlops*, which, among its other attributes, appears to undergo sudden, periodic torpor. The only evidence for this is the observation of Wood Jones[4] on a captive specimen, which he vividly described:

'It may be said to be like to the moles in its feverish restlessness. It evinces the most remarkable nervous activity. It makes endless tours around the confines of its cage, each peregrination being undertaken with characteristic energy and haste. It will search with feverish activity in each corner of its box, and regularly in each of the four corners turn a complete somersault in its enterprises. Suddenly it will discover something edible; the meal is accomplished with a maximum of speed; the performance is repeated and, in the case of a captive specimen I have observed, a handful of earth worms will be noisily accounted for in a short time. The meal having been digested at top speed, the animal will again start upon its rapid tour; but it has proceeded may be no more than a foot or two when, as suddenly as it awoke to activity, it is fast asleep. Even its sleep seems hurried. It breathes rapidly. It awakes with a start and is off again.'

'Apparently they need an extraordinary amount of food when active; but possibly when inactive they can sustain prolonged fasts. They seem to be animals which live either at fever pitch or to remain almost wholly quiescent as season and circumstance demand.'

Plate 10   The brown antechinus, *Antechinus stuartii*, from Queensland. (Stan and Kay Breeden)

Forest species

Antechinus stuartii

Species of *Antechinus* differ from the other small dasyurids in maintaining an even diurnal body temperature (Fig. 5.5) and never entering torpor. They live in the same forests as *Cercartetus* and, though predominantly ground dwellers, they can climb readily and have been trapped in large trees as high as 25 m. The particular interest in these species lies in their unusual life history rather than in their physiological responses to climatic extremes. *Antechinus stuartii* is best known, but what is known of the other species indicates a similar pattern.

Wood [292] followed the entire population of *Antechinus stuartii* on a 0·8 hectare plot of Queensland rainforest for 3 years, by repeated

recapture within the plot and on surrounding trap lines. The summary
of his results are presented in Fig. 5.6 and clearly show the annual
fluctuation in numbers of animals on the plot. Each year in January the
new cohort of young males and females appear in the traps soon after
they are weaned and are moving away from the parental nest. The adult
population at this time consists entirely of post-lactation females and no

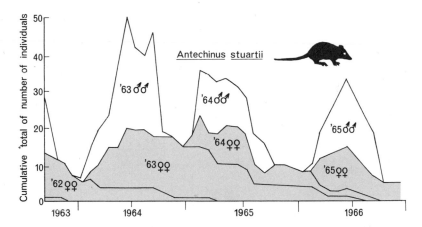

Fig. 5.6   Total number of *Antechinus stuartii* on 0·8 hectares of rain forest during
3 years, sub-divided to show the contribution of each annual cohort of young
males and young females. Note the complete disappearance of males in October
each year. (From Wood[292])

males at all. The reason for this can be seen in the total disappearance
of the males each October, whereas the females all survive into their
second and some into their third year. The peak in male numbers on the
study plot occurred each year in June and July and was due to an influx
of new unmarked males at this time. At the same time other males
previously marked on the study area were being caught on the outlying
trap lines, the two results indicating a greatly increased activity and enlarged range of the males at this time. The precipitate disappearance of
males was not due to a reversal of this process in October, since no males
were collected in any of the traps either within the plot or outside it.
Males caught just prior to disappearance showed loss of weight compared
to earlier captures and Wood concluded that the disappearance is due to
widespread death of the males at the end of September.

This conclusion had been foreshadowed by Woolley's[293] work on reproduction in *Antechinus stuartii* collected near Canberra. She studied a captive colony for several years and observed a very precise synchrony in the onset of sexual maturity in the young males and females, continuing into full reproductive activity (Fig. 5.7 and p. 38). The females, both one and two year animals, are monoestrous and come into oestrus in

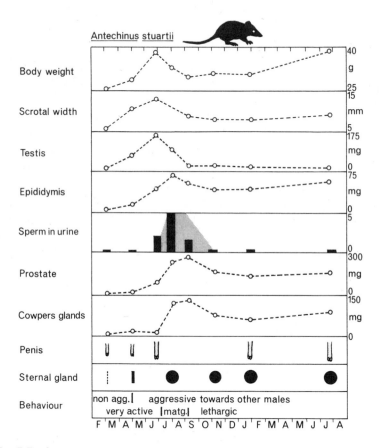

Fig. 5.7   Changes in the reproductive organs and associated parameters of male *Antechinus stuartii* from 5–22 months old. (From Woolley[293])

July-August. Oestrus lasts about 10 days and coincides with the peak of sexual development in the males. Copulation is frequent and prolonged[171] and the males are aggressive towards other males. By the end of August all the females are pregnant, and the males have passed into a general decline; spermatogenesis has ceased and the testes and accessory glands have regressed, the animals have lost much weight and their fur is falling out. With care they can be nursed through this time and will slowly regain weight and condition, but the seminiferous tubules and interstitial tissues of the testes never regenerate, so that they are quite sterile. Normally, however, the colony males have all died before the young are born to the females after the 30 day gestation, so that even if a female came into post-partum oestrus, which they do not, there would be no possibility of conception.

The rapid demise of the males cannot be ascribed to the sexual activity and fighting that prevail in July-August, because males isolated from other animals lose condition and die at the same time. It seems likely that there is an upwards shift in the metabolic rate of the males during the breeding season, and Woollard[292A] has observed that, although the food intake of males at this time increases, they change from being in positive nitrogen balance in April to a marked negative nitrogen balance of -15 mgN/g$^{0.75}$, whereas the females remain in positive nitrogen balance. This could account for the males' rapid deterioration in September but it leaves open the question of how it is controlled; is it an endocrine control exercised through the adrenals or is it a form of accelerated senility?

Apart from its ecological implications, the phenomenon may be a factor in the reproductive isolation between *Antechinus stuartii* and *Antechinus flavipes*, which are sympatric over much of their ranges and yet differ only slightly in morphological characters and indiscernibly in their ecology. *Antechinus flavipes*, however, breeds one month earlier than *Antechinus stuartii* near Canberra, and attempts to hybridize the two species failed because the respective males and females were too far out of synchrony.

CONCLUSION

These examples show that small marsupials, like small eutherians and small birds, possess a similar variety of adaptations to conserve energy and water. These are best seen in the smallest species and in those that live in environments at both extremes of the thermal range. There is, however, no direct parallel to the sudden demise of male *Antechinus* in the Eutheria but it is likely that it is related in some way to small size, since it is not a feature in the life history of the larger dasyurids, such as *Dasyurus viverrinus* or *Sarcophilus harrisii*, which are also monoestrous. The nearest eutherian parallel is to be found in the shrews, where both the males and females die at the close of the breeding season and only the juveniles overwinter to the next year.[49]

Other species of marsupial not mentioned may also show similar thermoregulatory adaptations upon closer study. The rabbit-eared bandicoot, *Macrotis lagotis*, is a small inhabitant of the central Australian desert, where it lives in deep burrows by day. One specimen examined showed a marked diurnal cycle in body temperature, whereas other peramelids from more temperate climates did not show such a fluctuation.[7]

The tropical didelphid, *Metachirus nudicaudatus*, also shows a diurnal temperature cycle, which may be associated with daytime torpor,[186] while Morrison and McNab[185] observed daily torpor in a single specimen of a Brazilian species of *Marmosa* during winter. This animal, which weighed 13 g, showed a diurnal fluctuation in T_B, similar to *Sminthopsis froggatti*. It was observed on many occasions to enter torpor spontaneously during the daytime but not at night. When it did so, its T_B fell from 34·7°C to 16°C and its oxygen consumption also fell to 14% of the basal level in the active state. Cold torpor or hibernation might also be expected to occur in the marsupials of the high Andes, especially the several species of Caenolestidae, which are small animals, less than 100 g, and live above 3000 m, but at present very little is known about them at all.

6

The Opossum—a successful Omnivore

It is often said that the opossum is North America's oldest mammal and its closest link with the Cretaceous mammals. This curious statement derives from its resemblance to the fossil didelphids of the North American Cretaceous, but, as we noted in Chapter 1, they entirely disappeared from North America by the Eocene; *Didelphis marsupialis* is a South American species, with an ancestry in that continent throughout the Tertiary, and it invaded North America after the land connection was established in the late Pliocene (Fig. 1.12). Nevertheless, the explicit motive given for a lot of the physiological studies undertaken on the opossum has been to compare this primitive species with more advanced ones, so as to understand the evolutionary origins of mammalian physiological processes. Since the first assumption does not stand, it is not surprising that many of these studies do not fulfil their authors' hopes; nevertheless they provide, inadvertently, much useful knowledge about the adaptations of this versatile and highly successful species.

ORIGINS AND PRESENT DISTRIBUTION

The several species of *Didelphis* have a wide distribution in South America,[117] extending as far south as latitude 40°. *Didelphis marsupialis* is generally restricted to lowlands, while the very similar *Didelphis albiventris* (*azarae*) occurs at higher altitudes (Fig. 6.1). Both are confined to forest habitats, being absent from the southern grasslands and the high Andean paramos. Most of the other Didelphidae occur in the same forest habitat where they occupy the niches of small to medium-sized carnivores. They include the four-eyed opossums, *Philander* and *Metachirus*,

Plate 11 The common opposum, *Didelphis marsupialis*, from Cali, Colombia (H. Tyndale-Biscoe, by permission of National Geographic Society).

and the arboreal, fruit-eating woolly opossum, *Caluromys,* as well as the much smaller murine opossums *Marmosa* (p. 219) and *Monodelphis*. The only aquatic marsupial genus, *Chironectes*, resembles the otter in habits and is also a forest inhabitant. A few marsupials are found in other habitats, such as the 'comadreja' of the Argentine pampas, *Lutreolina*, and the tiny fat-tailed *Dromiciops* of the bamboo thickets of the Chilean Andes. The three genera of Caenolestidae, *Caenolestes, Orolestes* and *Rhyncholestes* are small shrew-like animals with a long independent ancestry in South America (Fig. 1.12) but are now restricted to the high altitude scrub bordering the paramos or the equally cool climate of the islands of southern Chile.

The Pliocene union of South and North America provided a route for endemic species from each continent to move into the other. The marsupial species that spread north are all members of the forest habitat or Brazilian zone, while *Dromiciops, Lutreolina* and the Caenolestidae did not spread north. In Fig. 6.1 it can be seen that four of the genera that have spread varying distances northward have nevertheless remained within the northern extension of the same forest habitat. This is most striking for *Philander*, whose northern limit coincides precisely with the northern limit of tropical evergreen forest. The murine opossum, *Marmosa mitis*, extends slightly beyond this boundary up the western coast of Mexico to about 20°N.

By contrast, *Didelphis marsupialis* has spread, by natural means, throughout the temperate zones of eastern North America and the west coast of Mexico (Fig. 6.1). The desert country of north-west Mexico and New Mexico were a barrier to its natural spread west of the Rockies but, since its liberation in central California between 1870 and 1915, it has become established throughout the lowlands from southern California to southern British Columbia (Fig. 7.5). In eastern North America its northern limit at 40°N appears to be a climatic one since it coincides fairly closely with the −7°C January isotherm. During the past century there have been 25 records of opossums in southern Ontario but little evidence of an established population.[200] Sixteen of the 25 specimens were collected in the winter months and were found around those sites where the animals could most easily have crossed frozen rivers from Michigan or New York State. In Michigan[51] the northern limit is approximately the pine-hemlock ecotone between northern coniferous forest and the deciduous forest.

What in the ecology and in the physiology of this species enabled it to become so firmly established in the face of the probable competition from indigenous eutherian species during the Pleistocene, and what limits its further spread northwards?

Ecology

As with the brush possum in Australia, the opossum shows a cline in body weight from 1–2 kg in Texas up to 4–5 kg in the northern part of its range. By its size and its generalized dentition it is able to use a wide range of food from insects and earthworms to small vertebrates, as well as fruits and grasses. The remains of adult rabbits and larger mammals found in opossum stomachs are probably taken as carrion, since they are unlikely to catch and kill animals of this size. In Missouri the food preferences change seasonally, insects predominating in the warmer months and mammals in the cooler. It has a well developed taste perception,[267] similar to that of the Carnivora, and the lips and tongue are very well represented on the cerebral cortex (see p. 206). Opossums are nocturnal and live alone by day in dens or nests taken over from other species. Despite its fairly large size, the opossum is not long-lived, especially in the northern part of its range, but is very fecund, so that the population turnover is rapid.

In lower latitudes such as in Panama[87] and Nicaragua,[38] southern Texas and Florida, females come into oestrus first in early January and most births occur before the end of that month, but in California,[213]

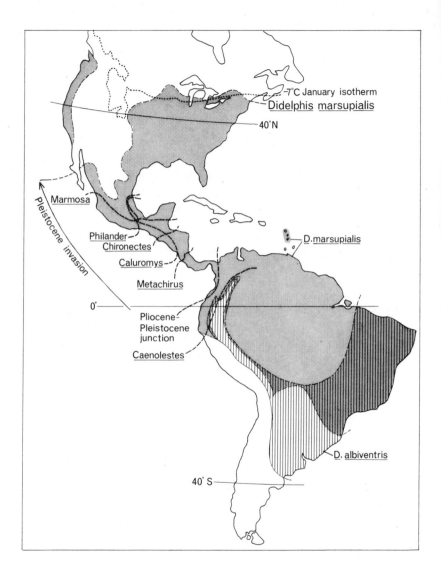

Fig. 6.1 Distribution of *Didelphis marsupialis* and *D. albiventris* in South America and the northern extension of the former species since the union of the Continents at the end of the Pliocene. From the position of the junction, in Colombia, it can also be seen how far the other genera of forest-dwelling marsupials have extended northwards since then. (From Hershkovitz[117] and Hall and Kelson[2])

Missouri[212] and Pennsylvania[38] the first births occur a month later. The number of young born often exceeds the accommodation of 13 teats in the pouch and further mortality of young occurs during the three months of pouch life,[226] so that the average number of young reared to independence is less than 8. If less than 3 young reach the pouch, lactation is not maintained and the female returns to oestrus in 15 days.[213] After weaning the first litter, also, the female will return to oestrus, so that a second peak of births occurs 3–4 months after the first. In Nicaragua a third litter may be reared, so that breeding is almost continuous throughout the year. Hartman[3] considered that opossums in Texas might also occasionally rear a third litter after weaning the second litter in July, but further north, where the second litter is weaned in August, all females enter a winter anoestrus. Furthermore, even a second litter is less frequent in Ohio[201] and Maryland.[161]

An opposite breeding season is seen in equivalent southern latitudes; around Rio de Janiero (20°S) two litters are reared by *Didelphis albiventris*, the first being born in July and the second in October.[123] Taken together these observations suggest that breeding may be initiated by a photoperiod signal such as we noted in *Macropus eugenii* (Fig. 2.2).

Since about 80% of the females produce young the potential fecundity per female per annum is about 12, which is almost ten times as great as in the phalangerids discussed in Chapter 4 (Fig. 4.5). In the few rather preliminary population studies undertaken on marked opossums the average life span from weaning has been estimated to be 1·3 years and the longevity about 4 years. Thus few opossums live beyond the summer following their birth. Females become sexually mature at 6 months, males at 8, but since, in the northern latitudes, anoestrus supervenes before they can realize this potential in the year of their birth, each animal in the main makes a contribution to only one breeding season. This very heavy mortality appears to be distributed across the whole population, although more so on the young animals newly weaned. However, as Lay[154] says . . . 'Opossums seem to lead a reckless life with little concern for the commonplace parasites, cuts, scratches, broken legs or tail, lost toes, ripped ears and broken teeth.'

Opossums are not territorial so that home ranges of individual animals overlap with those of others; neither are they social. Intra-specific competition may play some part in the exercise of juvenile mortality as it was shown to do in the brush possum, which is likewise a solitary animal. Predation by larger species is discounted by most observers, partly because the opossum is supposed to be unpalatable, and partly because of its distinctive behaviour pattern of feigning death. When seized suddenly by a dog or man, an opossum may become quite relaxed, as if dead; after release it will remain like this for several minutes, before cautiously raising its head and looking about. Failing to detect any sign of danger it will get to its feet and move off.

This behaviour can be evoked in young after weaning, but not before, and it is not evoked by sham attack nor by any stimulus except being grabbed and shaken.[94] In the absence of evidence that intra-specific competition or predation are the main causes of opossum mortality, environmental factors must be considered.

Thermoregulation

The opossum is a newcomer to the cool temperate climate of North America, from the more equable climate of the South American forests, so that its physiological capacities in this new environment may be limiting its further spread northwards as well as its longevity in lower latitudes.

In Panama, where the ambient temperature remains within 21°–32°C throughout the year, Morrison[186] found the body temperature of the

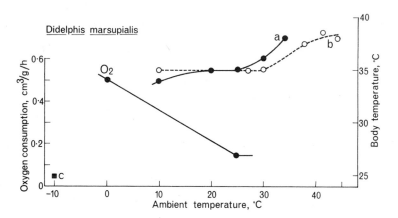

Fig. 6.2 A composite representation of thermoregulation in the opossum. (a. Values for T_B obtained by Morrison[186] at Panama; b. by Higginbotham and Koon[119] in Florida; c. T_B of cooled opossums by Nardone *et al*;[189] Oxygen consumption in laboratory acclimated animals from Brocke[51])

opossum to be $35 \cdot 5° \pm 1 \cdot 0$. The variation from the mean was irregular and in no way related to a diurnal fluctuation, such as we have seen in some dasyurids and *Marmosa*. In experimental conditions at Panama the opossum and *Metachirus* were able to thermoregulate between T_A 15°C and 30°C, but the T_B of both species rose at temperatures above 30°C

(Fig. 6.2a). *Metachirus* was unable to maintain T_B at 10°C, whereas *Didelphis* showed only a slight decrease at this temperature.

Responses to cold

At the higher latitudes of North America several separate studies have shown that *Didelphis* maintains a T_B of 34°–35°C at ambient temperatures down to 0°C,[119, 116A] (Fig. 6.2b). Wiseman and Hendrickson[290] observed that opossums during the winter in Iowa were active and emerged from dens to forage on days when the temperature was above −7°C, but were seldom active if it was colder than this and never if the temperature was below −12°C. The reason for this is made clear by studies of Nardone *et al*[189] who found that opossums held at $T_A − 10°C$

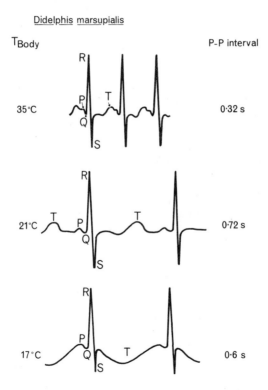

Fig. 6.3 Electrocardiograms of *Didelphis marsupialis* at normal body temperature and when the body had been cooled. (Redrawn from Nardone, Wilber and Musacchia[189])

could maintain a T_B of 34°C for only 20 minutes; thereafter it fell in one hour to 25°C (Fig. 6.2c) and in some specimens as low as 7°C. At these low temperatures the heart rate slowed down from a normal rate of 240 beats/min to 60 beats/min but, unlike the pigmy possum, *Cercartetus nanus* (p. 176), the electrocardiogram became altered in two particulars (Fig. 6.3); the decreased rate came about by a prolongation of the P–R, S–T, and T–P segments of the normal wave, while further cooling to below T_B 17°C affected the shape of the waves themselves, and inversion of the T wave. The first changes represent an alteration in the conduction rate of the cardio-accelerator nerve, while the subsequent effect at lower T_B is due to a direct inhibition of the cardiac conducting tissue in the bundle of His. These secondary effects are similar to those found in hypothermic dogs or rabbits, species that do not hibernate, but are not observed in species adapted to seasonal hibernation such as the ground squirrel, *Citellus*.

The opossum does not hibernate and, as these results show, it cannot remain active at ambient temperatures below − 10°C, so its habit of sheltering in dens, where the temperature is more tolerable, must presumably enable it to avoid winter extremes. Nevertheless, since it does not hibernate, as do other mammals in the same habitat, the opossum's energy requirements in winter are likely to be increased, in order to maintain T_B at 35°C. The thyroid gland is known to mediate responses to cold by enhanced secretion of thyroxine and in 57 opossums collected in Missouri, Bauman and Turner[31] determined that the thyroid secretion rate in adults was, indeed, significantly higher in winter and spring than in summer and autumn (Fig. 6.4). The rates in juvenile animals were

| Season | Adults | | Juveniles | |
	N	Mean T.S.R.	N	Mean T.S.R.
Winter (Dec.–Feb.)	18	1·92±0·47	3	3·58±1·90
Spring (Mar.–May)	16	1·94±0·37		
Summer (June–Aug.)	4	1·19±0·24	40	2·84±0·72
Autumn (Sept.–Nov.)	13	1·33±0·47	17	2·04±1·08

Fig. 6.4 Thyroid secretion rates, of *Didelphis marsupialis* in Missouri, determined as µg L-thyroxine/100 g body weight. (From Bauman and Turner[31])

consistently higher than the adult rates and their seasonal pattern was obscured by greater variation, due probably to the fact that the most rapid period of growth of young occurs in summer, after weaning which itself would be associated with enhanced thyroid activity. More direct measurements of the energy requirements of opossums in winter have been made by Brocke.[51] He measured the oxygen consumption of adult acclimated opossums, from Michigan at a range of T_A from 25°C, which is probably near to the thermoneutral zone, to T_A 0°C, a temperature at which the opossum can still thermoregulate.

At T_A 25°C the 3·5 kg opossums consumed 525 cm³ oxygen/hour (Fig. 6.2), which is equivalent to an energy expenditure of 264 kJ/day. This is lower than the normal range for other marsupials (Fig. 1.3), and considerably lower than for equivalent-sized eutherian species. At 0°C, however, the oxygen consumption was more than trebled and was equivalent to an energy consumption of 845 kJ/day. From published data on opossum winter diets, Brocke estimated that the energy intake from a winter night's foraging would be about 1380 kJ.

Thus, at the quite moderate winter temperature of 0°C, an opossum would be constrained to forage on two nights out of three in order to retain a positive energy balance, but, as we have seen in Fig. 6.1, opossums extend into country where the mean winter temperatures are considerably below this. Their survival in this environment can only be possible if the den temperature is near to the thermoneutral zone, and if they forage nearly every night. At night temperatures below −10°C this becomes impossible, so the need to forage on warmer nights becomes more imperative. In their study Wiseman and Hendrickson[290] observed that many of the opossums had frost damaged ears and tails. Others have observed that opossums in early spring are in very poor condition and that mortality in cage traps is highest at this time of year.

It would seem from this that the northern limit of the opossum is set at the −7°C January isotherm by the species' inability to maintain T_B below this temperature and, since it cannot conserve energy by hibernating, it is further limited by the proportion of winter nights on which foraging is possible. The opossum's lack of adaptations for hibernation provide independent physiological evidence for its geologically recent spread into North America, in contrast to the eutherian species of comparable size that have had all the length of the Tertiary in which to evolve these adaptations for extreme cold. One might predict therefore that the opossum will not spread any further north unless there is an amelioration of the climate and the −7°C isotherm shifts to a higher latitude.

Winter temperatures in the lower latitudes of the opossum's range in the eastern United States would also appear to be the main factor in the severe annual mortality, for even at 0°C, thermoregulation would require that the animal feed on most nights of the winter, or become progressively emaciated, and that it avoid extremes in a den. Food and den sites may thus be the limiting factors in winter survival and the main agents regulating the population.

If the opossum's ability to thermoregulate at low ambient temperatures has been an important factor in its invasion of North America, when compared to the failure of the other marsupials to do so, how does it cope with the other extreme of climate that it must have encountered as it moved out of the warm humid tropics to the hotter, drier continental summer climates to the north?

Responses to heat

When subjected to temperatures above 40°C the opossum's T_B rises to 38°C and can stabilize there for two hours [119] (Fig. 6.2b). Under these conditions the animals show similar behaviour to other marsupials we have considered; they become restless, start to pant, to salivate profusely and to lick saliva over various parts of the body, especially the legs and groin. At T_A above 44°C they lie on their sides or back with extended legs and some wave the hind legs with toes outspread. Attempts to detect sweating from different parts of the skin were negative as were measurements of differential changes in skin temperature, which would be indicative of vasodilation. Because of this Higginbotham and Koon [119] considered that the spreading of saliva over the body might be an important cooling device. They prevented one animal from spreading saliva but not from panting or salivation by means of anaesthesia. The T_B rose to 43°C and the animal died, so they concluded that spreading saliva was essential for thermoregulation. This conclusion has been accepted by later authors, but is far from an adequate demonstration of the importance of this behaviour, and the experiments on conscious quokkas [34] (p. 139) and possums [70] (p. 152) indicated an opposite conclusion in those species, and a greater importance to sweating and panting. Clearly the physiological responses of opossums to heat need to be examined more thoroughly, especially as the behaviour of spreading saliva is definitely evoked by thermal stimuli.

Hypothalamic control centre

The hypothalamus has been explored for a thermoregulatory centre.[218] By placing fine electrodes in different parts of the hypothalamus Roberts et al[218] were able to evoke a variety of behavioural responses with electrical stimulation. If they substituted heat stimulation, delivered via the same electrodes in the form of radio frequency current, however, no behavioural responses were elicited except when the electrodes were placed in the medial preoptic areas and anterior hypothalamus; warming of these areas to 38°C, while the rectal temperature of the opossum remained at 34·6°C, induced extensive licking of the body, limbs and face, together with panting and a sleep-like relaxation, but none of the non-thermoregulatory responses, such as mating, attack, or eating, that could be elicited by electrical stimulation in the same area. All these responses were the same as those observed in animals subjected to T_A above 44°C, when the T_B rises to 38°C, and they concluded that the localized warming stimulated temperature sensitive cells in the anterior hypothalamus, which control these behavioural patterns. The results also suggest that T_B is monitored by the blood temperature, rather than by afferent fibres from the skin or other peripheral regions of the body.

Petajan et al[199] tried to separate these factors in opossum thermoregulation by two surgical procedures; removal of the cerebral cortex, and transection of the spinal cord at the level of the upper thorax. Decorticate animals were able to maintain normal body temperature at high and low T_A, just as well as intact animals do, which confirmed a much earlier observation of the same effect by Rogers.[7] Moreover, Rogers had observed that thermoregulation failed if the mid-brain was transected behind the hypothalamus (Fig. 6.5), thus localizing the thermoregulatory centre to this region. Furthermore, in comparing pouch young opossums of 40 and 80 days, that is to say, before and after thermoregulation is achieved (Fig. 2.19), marked changes can be seen in the cellular differentiation of the preoptic and anterior hypothalamic areas.

Transection of the spinal cord destroys the afferent fibres that monitor skin temperatures. It also paralyses the skeletal muscles by destroying the motor neurons, so that shivering in response to cold will not occur posterior to the lesion. Transected animals were able to maintain T_B at ambient temperatures down to 10°C, after which they began to cool and their oxygen consumption also declined. The transected animals responded to low T_A by violent shivering of the anterior, innervated parts of the body; and this was equally pronounced when only the denervated portion was cooled by immersion in cold water. This showed that the lowered blood temperature from the posterior end was the main factor stimulating the hypothalamic centres and that lowering of the skin temperature is less important. Also the results show the thermogenic importance of shivering to the opossum, since it was unable to produce sufficient heat from the anterior end alone when the difference between T_B and T_A became greater than 24°C.

These several observations thus indicate that the opossum is a true homeothermic mammal, with a hypothalamic centre that is sensitive to blood temperature and that mediates behavioural and physiological responses to maintain a stable body temperature within a wide range of ambient temperatures, from about $-7°C$ to $44°C$. In common with other marsupials the normal body temperature is about $3°C$ lower than the majority of eutherian species observed, but this is not associated, in this or other marsupial species, with less effective regulation of body temperature. It is however, associated with differences in several blood parameters directly involved in oxidative metabolism.

Blood

The blood volume of adult opossums is $56 cm^3/kg$, which is the same as the brush possum (Fig. 4.4), and within the lower range of a wide variety of eutherian species examined by Burke.[56] The haematocrit, or red cell volume, is lower than in most eutherians, being 35%, and the haemoglobin concentration at $11 g/100 cm^3$ is also lower than most eutherian species and as low as quokkas in summer anaemia (p. 138).

Kidney function

In other functions the opossum displays adaptations comparable to eutherian species of similar habits. The opossum kidney has been studied in some detail and the results can be compared to those previously discussed on the brush possum and to a lesser extent on the macropods and *Dasycercus* (Fig. 3.18).

The functional capacity of the opossum kidney is remarkably similar to that of *Trichosurus vulpecula*. During water deprivation it can maintain the plasma electrolytes at the normal concentration of $318 m$ osmole/dm^3 and can produce a urine $4·7$ times as concentrated.[206] This is a higher concentrating ability than is achieved by the pig or man but not as high as the dog or the macropods and much less than *Dasycercus cristicauda*. The osmolarity is the sum of the electrolyte concentration and the urea concentration, so that for an animal whose diet includes much protein, a large contribution to the solutes of the urine will be urea from protein catabolism. It is immaterial whether the protein is taken in as the flesh of prey by a carnivore, or is derived from symbiotic bacteria in the stomach, so that carnivores and ruminants resemble each other in this respect more than either resembles the omnivores or non-ruminant herbivores. For the ruminant the high urea production can be re-utilized in further bacterial synthesis, as we have noted earlier, but in a carnivore the urea must be excreted through the kidney. If water is abundant no competition arises between the discharge of urea and inorganic electrolytes, especially K^+, but how is this resolved under conditions of dehydration? Does the excretion of urea inhibit the excretion of other electrolytes when the kidney is functioning at its maximum concentrating ability or are the two independent?

Plakke and Pfieffer[206] have examined this in the opossum by depriving animals of water after diets of high and low protein and high and low electrolyte content in different combinations. The results indicated that neither elevation of the urine electrolyte concentration nor the urine urea concentration depressed the other and that, in fact, electrolyte excretion is enhanced by increased urea concentration. Several eutherian species exhibit this response to urea, including man, dog, *Dipodomys*, rabbit and musk rat, whereas the kidney of the beaver and pig do not exhibit this ability to concentrate urea independently of electrolytes. Plakke and Pfieffer[202, 204] have correlated this ability with particular anatomical features common to the opossum and the eutherian species mentioned. As we have already noted, species with a good renal concentrating ability have a relatively thick medulla and nephrons with long loops of Henle. As well as this the vascular loops (vasa recta) that dip into the medulla, anastomose into vascular plexuses at several levels so that there is a distinct zonation in the medullary blood supply. Furthermore, the pelvis of the kidney in these species is not a simple expansion of the ureter as it is in the pig and beaver, but is an extensive chamber whose wall is thrown into elaborate folds that penetrate deep into the inner zone of the medulla. Plakke and Pfieffer consider that these anatomical features may provide a more effective counter-current exchange system between the fluids in the nephrons, the vasa recta and the folds of the pelvis and thus make possible the greater urea concentrating capacity that these species display. The kidney of *Trichosurus* also shows zonation of the medulla as does the kidney of the tammar, *Macropus eugenii*, so it is possible that the anatomy is similar in these marsupials too. Also as noted before, stop-flow analysis of the possum kidney showed that considerable re-absorption takes place in the inner zone of the medulla.

Notwithstanding that the opossum shows these anatomical adaptations for enhanced urine concentration, its abilities in this respect are moderate by comparison with species adapted to arid environments, and it compares more closely to the brush possum which is also a forest dweller (Fig. 3.18). Perhaps in this respect, as in the opossum's inability to hibernate, its ancestry has imposed a limitation on its distribution in North America, since it does not penetrate into those parts which have a low rainfall in western North America or western and southern South America. Both these regions are comparable to the habitats of the desert macropods and *Dasycercus cristicauda* in Australia, all of which have better renal concentrating ability than the opossum.

Endocrine control of kidney function

The endocrine control of kidney function in the opossum has attracted interest since the unexpected results of Britton and Silvette's experiments

40 years ago on the effects of adrenalectomy.[7] Although they removed both adrenals in one operation, their animals survived for much longer than eutherian species did. The animals did not show the decline in plasma Na^+ and rise of plasma K^+ normally associated with adrenal insufficiency but they did show lowered blood sugar and liver glycogen concentrations, which eventually caused their death. These results were confirmed by Hartman, Smith and Lewis,[7] who found that if the opossums were adrenalectomized in a two stage operation their survival was even better and they would live several months if not stressed. It was concluded that sodium retention in the opossum is less dependent on mineralo-corticoid hormones than it is in eutherian species. The subsequent discovery that other marsupials are not so tolerant of adrenal loss, and the much greater understanding of the role of aldosterone and its control by secretions of the juxtaglomerular cells of the kidney, has re-awakened interest in the opossum.[137]

Blood taken from the adrenal vein contained aldosterone, cortisol and corticosterone but, as in other marsupials measured, cortisol is the most abundant. Secretion of all three hormones fell after hypophysectomy or treatment with dexamethazone, a drug that inhibits anterior pituitary activity. The decrease was least for aldosterone ($\frac{1}{2}$) and greatest for cortisol ($\frac{1}{10}$). Infusions of sheep adrenocorticotrophic hormone (ACTH) immediately enhanced the secretion of all three steroids 3–4 fold. Infusion of opossum kidney extract containing renin enhanced secretion of aldosterone and to a small extent of cortisol also. In other species renin is secreted by the juxtaglomerular (JG) cells in response to reduced blood pressure in the renal arterioles and possibly to reduced plasma Na^+; it catalyses the transformation of a plasma constituent, angiotensinogen, to angiotensin, which stimulates a rise in blood pressure and secretion of aldosterone from the glomerulosa zone of the adrenal cortex. It is thus interesting that sodium depleted opossums showed an increased activity of juxtaglomerular cells and also that opossum renin stimulated increased blood pressure and aldosterone secretion.

The evidence that a renin-angiotensin-aldosterone system occurs in the opossum is difficult to reconcile with the earlier conclusion that the adrenal cortex is not essential for salt balance in the opossum. This can probably only be resolved by repeating the earlier work and monitoring the animals for circulating steroids after adrenalectomy. This would detect any regeneration and determine whether there are other sources of adrenocortical hormones in the opossum, which respond to ACTH and renin. It seems most unlikely that such an elaborate kidney and such considerable renal concentrating ability as the opossum has, is not dependent upon aldosterone or some other cortico-steroid for its proper functioning.

STRUCTURE AND FUNCTION OF THE BRAIN

The opossum is said to be a very stupid creature, to display no social behaviour or play and to be incapable of being tamed or of discriminatory learning.

Attempts to relate this to brain structure of the opossum have not been successful, although a great deal of work has been done on the gross and microscopic anatomy of the opossum brain and on determining the connections between the several parts. Because the opossum is the best worked marsupial, it seems appropriate to treat the brain in this chapter and refer to other species, especially *Trichosurus vulpecula*, where relevant.

The vertebrate brain can be thought of in the first instance as a closed tube, the walls of which comprise four columns of neural tissue; the two dorsal columns contain the sensory components derived from the incoming or afferent nerve fibres and the two ventral columns contain the motor components or efferent fibres. The brain tube is so constituted that a fore, mid and hind brain can be distinguished, the fore brain being again divisible into an anterior telencephalon and posterior diencephalon and the hind brain being divisible into an anterior metencephalon and posterior myelencephalon. In mammals this structure, which is termed the brain stem, is overlaid by the enormous enlargement of the roof of the telencephalon, to form the paired cerebral hemispheres, and of the metencephalon to form the cerebellum.

Each of these structures is linked to the brain stem and to each other by large tracts of nerve fibres, known as commissures. Because most of the component fibres are surrounded by myelin sheaths, the commissures appear white in gross dissection, whereas the parts where cell bodies of the neurons are concentrated appear grey; where sufficiently concentrated these parts have a discrete appearance and are known as nuclei and it is at these nuclei that fibres from different parts of the brain interconnect by synaptic junctions and new fibre tracts arise to other parts, so that nuclei can be thought of as relay stations in brain activity.

Co-ordination of varying complexity is achieved in such relay stations at all levels of the brain, but in mammals the cerebral hemispheres and cerebellum have become paramount over brain stem centres and most study has been directed to determining their connections to brain stem nuclei. Three approaches have been most commonly employed. The first exploits the observation that the distal portion of an axon dies if its cell body is destroyed and the degeneration can be stained selectively; thus, by making lesions in particular parts of the brain, all the outgoing fibres can subsequently be traced to the next synapse, by means of stained serial sections of the entire brain. The second procedure is to stimulate

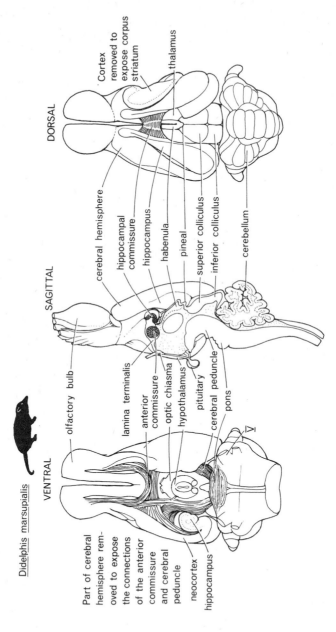

Fig. 6.5 Gross anatomy of the opossum brain to show the main parts and commissures. (From Loo[162])

electrically particular parts of the brain, usually the cerebral cortex, and observe the responses in the body, or monitor the electrical responses in other parts of the brain. This is especially useful for studying motor function. The third method is to monitor electrical activity in brain tissue that may have been evoked by stimulation of sense organs or by electrical stimulation of other parts of the brain. By measurements of the elapsed time after the stimulation it is possible to estimate how many synaptic junctions are involved in the particular relay, since proportionately more time is spent exciting a synapse than in transmitting an impulse along an axon. An impulse can be directed along an axon towards the cell body but such an antidromic impulse will not pass across a synapse in this direction, so antidromic excitation can be used to determine the origin of axons, in contrast to degeneration studies that disclose their endings.

Anatomy

The gross morphology and main commissures of the opossum brain are shown in Fig. 6.5. Far greater detail can be obtained from the work

Marsupial   Bat   Primate

Didelphis marsupialis   Eptesicus fuscus   Saimiri squireus

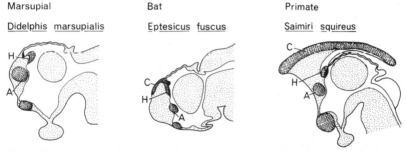

Fig. 6.6 Sagittal sections through the brain stem of the opossum and two eutherian mammals, to show the disposition of the corpus callosum (C) in relation to the hippocampal (H) and anterior (A) commissures. (From Akert, Potter and Anderson[11])

of Loo[162] and Bodian.[42] In the fore brain two transverse commissures link the cerebral hemispheres, the large anterior commissure and the smaller hippocampal commissure. In the Eutheria a third commissure, the corpus callosum (Fig. 6.6), links the two cerebral cortices but it is absent from marsupials and monotremes and is insignificant in some eutherian species, such as bats. Its absence from marsupials, however, has been considered the reason for their lesser powers of integration.

The main commissures linking the cerebral hemisphere with the brain stem and spinal cord are the paired cerebral peduncles, which arise in the corpus striatum, underlying the cortex, and pass along the ventral floor of the midbrain to terminate in the ventral medulla. Lying ventrally across these commissures is the pons varolli, which links the two sides of the cerebellum and is unique to mammals. A small distinction here separates the Monotremata from other mammals in that cranial nerve V emerges in front of the pons, not behind it, as in the Marsupialia and Eutheria. Thus in gross structure the brains of all mammals are quite similar to each other and quite different from other vertebrates. The questions that arise in the present context are how the marsupial brain differs in fine structure or function from the eutherian and at what stage in pouch life it differentiates into a functional mammalian brain.

Motor functions of the cerebral cortex

Lende[158] has made the most detailed study of the motor cortex in *Didelphis*, by electrical stimulation of the exposed cerebral hemispheres, and has shown (Fig. 6.7) that all parts of the body are represented on

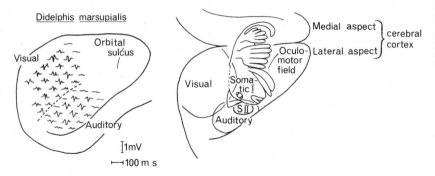

Fig. 6.7 Surface view of right cerebral hemisphere of opossum with sensory and motor representation superimposed. On left the results in one animal showing the patterns of evoked potentials to light flash and to click. The largest potentials each side of the line indicate the areas of maximum response to visual and auditory stimuli respectively. On the right the simulacrum represents both the somatic sensory and motor areas, while eye movements are induced in the pre-orbital region. (After Lende[158])

discrete parts of the cortex; the anterior parts of the body and especially the facial regions and fore paws have the largest representation, whereas the hind limb and tail have the least. Indeed, in earlier studies it was believed that these were unrepresented on the cortex altogether and that, therefore, their activity was under thalamic control only and this was taken as evidence of the opossum's primitive nature. Similar patterns have been shown in *Trichosurus* and other marsupials.[7] Marsupials differ from Eutheria in lacking a second motor representation, anterior to the orbital sulcus, in the prefrontal area, but in both groups of mammals the strongest representation is contra-lateral. That is to say, stimulation of the left cortex elicits a movement on the opposite side of the body and only at greater stimulation is the ipsilateral side influenced. The reason for this is seen by tracing the route of the main efferent fibres. The cortical discharge begins in large neurons whose axons pass via the corpus striatum and the cerebral peduncles of the same side, to the ventral columns of the medulla in the pyramidal tract. There they cross to the opposite column, or decussate, and finally make synaptic connection with secondary neurons in the neck region of the spinal cord. These long fibres have been demonstrated by degeneration, after making lesions in the cortex of *Didelphis*;[32] and in *Trichosurus*, stimulation of pyramidal tract fibres in the medulla evoked antidromic responses in the intact cortex.[7] The pyramidal tract is only found in mammalian brains and has evolved, presumably, in close conjunction with the development of the cerebral hemispheres. The fibres are myelenated and therefore transmit impulses fast, thus increasing the direct control of the fore brain over motor functions. The second important connection is between the motor cortex and the cerebellum, since the latter organ is profoundly involved in the control of fine movements concerned with the animal's orientation in space. Again degeneration studies following cortical lesions[174] have shown that all the areas of the cortex involved in motor activity have direct connection with the basal nuclei of the metencephalon and terminate by synapse in the pons varolli.

The essential role of the cortex in motor activity can be explored more fully by its partial or total removal.[7] When the cortex on one side was removed without damaging the underlying corpus striatum, opossums showed only temporary paralysis, but damage to the corpus striatum resulted in permanent paralysis of the opposite side of the body. Conversely, total removal of the cortex, without damage to the corpus striatum, resulted in prolonged activity. Taken together, these observations suggest that the main control of motor function resides in the thalamus and that the cerebral cortex exerts a tonic control of its activity; this is a less profound control than that exercised by the cerebral cortex of advanced Eutheria, but results, similar to those on the opossum, from other marsupials would be required before it could be concluded that this is generally true for marsupials.

When the entire fore brain was excluded from exercising motor control in the opossum, by transecting the midbrain between the superior and inferior colliculi (Fig. 6.5), the animals showed decerebrate rigidity with spastic extension of all the limbs, elevation of the head, and an inability to move forward. This is interpreted as being due to exaggerated contraction of the anti-gravity or postural muscles, controlled by the cerebellum, without the tonic inhibition exercised by the fore brain.

Pouch young opossums, less than 50 days old, are unaffected by transection of the midbrain, or by cerebral lesions. Neither can movement be evoked by electrical stimulation of the cerebral cortex. The pyramidal fibres grow down from the cortex to the medulla by the 41st day but do not become myelenated until later. Thus, the complex movements executed by the newborn opossum in travelling to the pouch are not under the control of the cerebral hemispheres or the fore brain; the cerebellum may be involved or they may be spinal reflex movements.

Sensory representation on the cerebral cortex

Lende[157] has also made the most thorough study of the sensory cortex, by detecting evoked potentials from the exposed surface of the cerebral hemisphere to visual, auditory and tactile stimulation of anaesthetized opossums (Fig. 6.7). The areas for visual and auditory stimuli overlap slightly but the main area responsive to touch does not overlap either of these, and can be represented as a simulacrum of the animal. The distortion is due to the relative importance of the different parts of the body for tactile information. Clearly the nose, lips and tongue are important as are the fore paws. In a comparable pattern for the rat the paws occupy a smaller area but in the monkey a larger area. A second smaller, somatic area II, completely overlaps the auditory area. It is evident that the sensory and motor areas of the cortex overlap completely, and coincidentally. A similar but less precise overlap of these areas of the cortex has been shown in *Trichosurus vulpecula*,[7] and it is general in Eutheria as well; it reflects the role of the cerebral cortex in co-ordinating incoming sensory information and outgoing motor impulses.

The afferent pathways to the cortex have been traced by sequential degeneration studies from lesions in brain stem nuclei. These show that *Didelphis* and *Trichosurus* conform to the normal mammalian pattern; incoming fibres from the sense organs and spinal cord synapse in the thalamic nuclei and thence other neurons project to the cortex. As well as these, a large component of the incoming fibres are from the opposite cerebral hemisphere via the anterior and hippocampal commissures;[84] when these commissures were transected degenerating axons were traced to all parts of the cortex on both sides, so that it appears that the absence of the corpus callosum—the additional eutherian commissure—does not deny the opossum close connection of the two hemispheres.

This is borne out in a study by Nelson and Lende,[190] who monitored evoked potentials on one cerebral hemisphere resulting from electrical stimulation of the same area on the opposite hemisphere. In intact animals a primary response was evoked within 15 msecs and a secondary response within 80 msecs. The fast response, probably monosynaptic, was abolished after the anterior and hippocampal commissures were cut, but the slower response survived these cuts as well as a mid-brain transection. It is concluded from this that the main interconnection is by the two large commissures but other, slower, connections must occur via thalamic fibres and several synaptic links.

The brain stem

Notwithstanding the dominant role of the cerebral hemispheres in neural integration in the opossum, the older brain stem centres are profoundly important. In the diencephalon the lateral components are the main relay stations to the cerebral cortex, while the ventral component is the hypothalamus. This, as we have already seen, contains the centre for thermoregulation, but as in other mammals it also mediates other behaviour patterns. These have been explored by electrical stimulation[219] with electrodes placed in various parts of the anterior hypothalamus and preoptic area. The responses evoked could be divided into two kinds. Grooming, investigatory, and threat responses could be evoked directly by the electrical stimulus. Other, goal directed patterns, such as the various components of attack, eating and of male sexual activity, only occurred if the objects, such as a rat or an ovariectomized female opossum, were present, indicating that the behaviour was facilitated, rather than evoked, by stimulating the hypothalamus.

CONCLUSION

This brief survey shows that the functional and morphological attributes of the opossum brain closely resemble other mammals and no substantial differences can be readily discerned. If correlations do exist between the opossum's brain structure and its reputedly low intelligence they are not detectable by present methods of investigation and must await future advances in neurophysiology. By the same token, the opossum's reputation for stupidity needs to be examined more rigorously than heretofor, to determine precisely in what ways it is inferior to other mammals with similar life patterns.

In its other physiological functions, as in its brain, the opossum does not display the primitive features anticipated by its close resemblance, in teeth and skeleton, to Cretaceous marsupials; rather, it shows many adaptations that have their parallel in eutherian species of similar habits. It remains a worthy object for research, not for the reasons often given, that it is a living Cretaceous fossil, but because, like other marsupials, it shows how a distinct and separate mammalian organization has evolved parallel adaptations in response to the same environmental imperatives.

7

Marsupials and Man

PLEISTOCENE EXTINCTION

Throughout this book the emphasis has been on how marsupials are severally adapted to the environments they encounter, so it is appropriate to conclude by considering how they have responded to the latest and most preponderant factor in their environment today. Until very recently it was generally believed that Palaeolithic man made little impact upon the other mammals he encountered and that he was a child of the environment as much as they were. On this view, the widespread extinction of many genera of mammals and birds at the close of the Pleistocene resulted from the increasing aridity at the end of the last Ice Age, 12 000 years ago, and only since the rise of civilization has man become an important factor in the extinction of animals.

Several awkward facts obtund this view of Pleistocene extinction. First, very few mammals became extinct at the close of the three previous Ice Ages, and the late Pleistocene extinctions differ from those of the Tertiary, in that the species affected were predominantly large animals and were not replaced by later evolved ones, so that their niches have since remained unoccupied. This was most dramatically shown by the spread of the horse across North America after its reintroduction by Cortes in 1519. Not so the species less than 50 kg body weight, which remained abundant and diverse, except on the West Indian Islands and Madagascar, where they suffered a similar eclipse to the large species on the continents. Second, the times of extinction can now be determined precisely by measuring the relative amount of the radioactive isotope of carbon (C^{14}) in organic matter. Since it decays at a constant rate to C^{12}, the time at which it constituted the same proportion as in the CO_2 of the air can be calculated, and this indicates the time that the carbon in the organic matter was being actively exchanged with the atmosphere and hence the time that the animal or plant was alive. These observations have made it clear that the Pleistocene extinction in, for example, North America took place within a very short period, less than 1000 years; similarly, the extinction of the great moas of New Zealand was rapid, and so too the lemurs and ratite birds of Madagascar. Furthermore, while the extinctions in different parts of the world show the same pattern, they did not occur at the same time; only those of North America and Europe occurred at the end of the Ice Age, others occurred later and some earlier. The one event common to every extinction was the appearance of palaeolithic man at about the same time.

These observations led Martin[177] to propose that late palaeolithic man was the main agent causing the Pleistocene extinction, either by his hunting or, indirectly, by the alteration of the habitat wrought by fire. There are still a number of difficulties in this fascinating hypothesis, the most puzzling being the survival of far more genera of mammals in Africa and Eurasia, where man has been longest, than in the New World. Martin considers that a major extinction did occur in Africa much earlier, about 40 000 years ago, coinciding with the development of the advanced Acheulian culture, which spread north and east from Africa. One suggestion is that extinction was less dramatic on these continents because man and mammals evolved together, but when the Ice retreated and man crossed the Bering Straits about 12 000 years ago, the large animals had not evolved an escape response to this novel predator and fell easy prey, as in later times did the seals and penguins of the Sub-Antarctic islands.

In South America extinction of the large Pleistocene mammals occurred about a millenium later than in the north and, though not so fully known, was even more profound. As in the north the smaller species were unaffected. This included the marsupials, which by the Pleistocene were all small animals, the large carnivorous borhyaenids having become extinct much earlier. Conversely, one might speculate that the northern extension of range that the opossum underwent was made possible by the ecological disturbances consequent upon the rapid extinction of the large herbivores; it has been suggested that their removal must have been followed by an increase in the smaller vertebrates and invertebrates, which the omnivorous opossum could readily exploit. This notion could be tested if the northward spread of the opossum can be adequately dated and can be shown to post-date the major mammalian extinction of 11 000 B.P.

AUSTRALIAN MARSUPIALS AND PALAEOLITHIC MAN

In Australia the large herbivorous marsupials and ratite birds became extinct in the late Pleistocene, several millennia after the arrival of man on the continent. The time of extinction is not nearly so well known as in North America, so that the role of man cannot be readily discerned. As well, the broad period of extinction overlaps the 'Great arid' period of climatic change in southern Australia, when the distributions of many species of marsupial shifted, so that local extinctions occurred although the species survived elsewhere.

Figure 7.1 summarizes some of the precise dates, based on C^{14} measurements, for human occupation and for the remains of 13 marsupials and monotreme genera now extinct on the mainland of Australia. The sites are among those shown as Quaternary on Fig. 1.13.

Two distinct cultural phases are now recognized in prehistoric Australia.[188] The first phase is distinguished by unifacially worked pebbles and by the use of flake tools and the parent core. These were heavy, hand-grasped tools similar to the early to mid-Palaeolithic culture of other lands. The tools occur in the lowest archaeological strata in southern Australia and on Flinders Island, Kangaroo Island (20 and 22 on Fig. 3.15) and Tasmania. These people reached the islands before rising sea levels cut them off from the mainland at the end of the last Ice Age 11 000 B.P. They survived on Tasmania with essentially the same culture, until exterminated by the Europeans a century ago, but had died out on the two smaller islands before this time. They did not have dogs, nor have remains of dogs been found at sites on Kangaroo Island, from which it is concluded that the dingo arrived after the separation of Tasmania.

On the mainland the flake and core culture evolved slowly with little sign of innovation. It was superseded from the north by a much more advanced stone culture, characterized by finely worked pirri and bondi points, which were hafted to wooden handles, and by mills for grinding seed. In other parts of the world this microlith culture is associated with domestic dog, and it is probable that the two aspects of this new culture entered Australia together between 7000 and 5000 B.P.

Early Palaeolithic man lived on Borneo 32 600 years ago and the earliest C^{14} date for human remains in Australia is somewhat later, though some consider that flake and core culture in South Australia may be this ancient. Mammoth Cave in south Western Australia is more than 31 000 years old and contains the remains of five extinct marsupial species (Fig. 7.1), some of which show evidence of charring,[179] but it is not clear whether this was human activity or caused by natural fires. There is less doubt, however, of the earliest deposits around the shores of Lake Menindee in New South Wales. Two samples of charcoal from

Date B.P. by C14	Locality	Human culture	Canis dingo	Marsupial Genera												
				Sarcophilus	Thylacinus	Thylacoleo	Diprotodon	Nototherium	Zygomaturus	Macropus ferragus	Protemnodon	Sthenurus	Procoptodon	Propleopus	Phascolonus	Zaglossus
32 630 ± 700	Borneo	early Palaeolithic														
31 500	Mammoth Cave, W.A.	charred bones		+	+	+	+		+	+	+	+	+	+	+	+
26 300 ± 1500 / 18 800 ± 800	Lake Menindee, N.S.W. (layer B)	charcoal, flake and core		+	+	+			+			+	+			
22 700 ± 700	Malangangerr, N.T.	flake and core tools														
14 000 ± 250	Boolunda Creek, S.A.	—						+								
13 725 ± 350	Lake Colungalac, Vic.	?					+				+		+			
11 000		(Tasmania isolated)														
8600 ± 300 / 7450 ± 270	Mt. Burr, S.A.	flake industry	+													
4770 ± 150	Tyimede 2, N.T.	first microlith pirri points		+												
3120 ± 100	Padypadiy, N.T.	—	+		+											
2590 ± 350	Nullarbor Cave, W.A.	—			+											
538 ± 200	Koroit beach, Vic.	midden		+												

Fig. 7.1 Chronological table of important radio carbon dates for human culture and extinct marsupials in Australia. (From Merrilees,[179] Mulvaney,[188] and Tedford[269])

there have been dated at 26 300 and 18 800 years B.P., and are associated with flake and core artifacts, as well as the bones of 10 extinct large mammals and 13 smaller genera, still extant today.[269] Many of these bones show evidence of having been handled by men, being broken or charred. In South Australia and Victoria seven of the extinct genera have been recovered from sites dated to 13 725 years B.P., and the most recent specimen of *Diprotodon* has been dated 11 100 years B.P.

What is clear from these dates is that early Palaeolithic man had a long association in Australia with the large diprotodontids, macropodids and marsupial carnivores lasting for about 20 000 years, and that he probably hunted them. Rock carvings[30] in the foothills of the Flinders ranges, 250 km from Lake Menindee, depict the footprints of *Diprotodon* and the extinct giant emu, *Genyornis*, among other carvings of emu and kangaroo footprints of normal size, and of human figures, which indicates a familiarity with these extinct animals. Absent from these carvings are dog footprints, although these are common in other carvings, where the extinct species are not depicted; no remains of dog have been found in any of the sites so far mentioned and it seems unlikely that these people had dogs.

Overlying the early strata at Lake Menindee is a much later bed containing point and blade artifacts of the later microlith culture. There is no good date for this site but two doubtful values are about 6000 B.P. The site is remarkable, however, because it contains none of the extinct species from the earlier strata, and, with two exceptions, they do not occur in any subsequent sites investigated.

The exceptions are the marsupial tiger, *Thylacinus cynocephalus*, and the Tasmanian devil, *Sarcophilus harrisii*, the largest of the carnivorous marsupials apart from *Thylacoleo*, which disappeared with the Diprotodontidae. *Thylacinus* appears in a number of other sites in southern Australia, the most recent being a desiccated corpse found in a cave on the Nullarbor plain, which has been dated at 2590 B.P.[6] *Sarcophilus* survived on the mainland for much longer, the latest specimen from Victoria being 528 B.P. Both species still exist in Tasmania, to which the later microlith cultures did not penetrate and the dingo did not reach.

Thus, during the period between 10 000 and 6000 B.P. there was a major cultural change in the human populations from heavy coarse stone implements, to fine, well-made microliths and the first appearance of stone mills for grinding grass seeds. Most of the large animals had disappeared by then and the dog had made its entry. The earliest good date for this in South Australia is 7–8000 B.P.[188] (Fig. 7.1). Since it has not been found in other sites earlier than this it is most probably near the time of its arrival. Calaby has argued, on the analogy of the fox, which crossed Australia in less than 70 years, that the dingo would have become widespread within a century of its introduction, and may have preceded the men who brought it. If this is correct, the dingo was not associated with the decline of the large animals during the earlier times. It has been suggested that it had an effect on the smaller species and on the extinction of *Thylacinus* and *Sarcophilus*. There is no good evidence of the smaller species suffering extinction, although many show alteration of range, and *Canis* was contemporary with *Thylacinus* for 5000 years and with *Sarcophilus* for 7000.

The hypothesis that man exterminated the large marsupials and birds in Australia by 11 000 B.P. rests on slender evidence, but there are sufficient similarities to the North American pattern for it to be entertained as a useful hypothesis. The species that disappeared were the large ones and were not replaced by another array of grazing and browsing species, while the smaller species persisted. Human artifacts changed from large adzes and scrapers suitable for flensing, to microliths suitable for arming spears to use on smaller game.

Against this hypothesis is the lack of evidence of a very rapid extinction, such as occurred in North America, as well as the good evidence of marked climatic change, reflected not only in soils and vegetation but in the shifts in distribution southwards of marsupials that formerly occurred further north; some got left behind on islands, such as *Lagostrophus fasciatus* on the Shark Bay islands (Nos. 5–7 on Fig. 3.15); others became locally extinct, such as the wombat *Lasiorhinus*, which no longer occurs near Lake Menindee, as it did 26 000 years ago. If climatic changes at the close of the last Ice Age did indeed initiate the decline of the larger species, the arrival of Palaeolithic man, with his habit of firing large tracts of land, may have precipitated their end; the dingo, often thought to have been an important agent, seems to have been far less so than habitat change, for when Europeans first encountered Australia 300–200 years B.P., the marsupial fauna was abundant and diverse.

AUSTRALIAN MARSUPIALS AND EUROPEAN MAN

Although the first record of an Australasian marsupial was made by Torres' party when they landed on New Guinea in 1606, the first accurate description was given by Pelsaert, when he returned to the Abrolhos islands to rescue the survivors of his ill-fated ship Batavia. He described the tammar on West Wallaby Island and was astounded by its mode of generation. Subsequently other Dutch and English navigators made landfall on the islands and coast of Western Australia and described other species, but the main exploration of the continent and concurrent description of the fauna began after the Penal settlement at Botany Bay was established.

Status at the time of settlement

The period from 1788 until 1850 saw the establishment of settlements around the eastern and southern coasts of Australia and on Tasmania, and the expansion inland by the sheep men to occupy the plains of the Murray-Darling basin. As well, the great expeditions set out from these settlements to explore the coastline and to cross the continent. From all of this activity increasing numbers of marsupial and other species came into the hands of naturalists and were described and recorded. This culminated in the work of John Gould and John Gilbert, his indefatigable collector between 1838–1845, with the publication of Gould's 'Mammals of Australia' in 1863. Until then European influence upon the marsupial fauna had been slight but after 1850 the close settlement of land by selectors was actively promoted by Government and led to the wholesale clearance of the lightly timbered land and the intensive stocking of the grasslands. At the same time the rabbit became established and by 1880 had occupied all the settled land as far north as Queensland.

Very few species of marsupial have been discovered since Gould's publication, but many have since disappeared, so that his work, and Thomas' Catalogue[270] represent the datum point for our knowledge of the marsupial fauna at the time of European settlement. Public awareness that the marsupials were disappearing from settled country probably began at the beginning of this century and was expressed most cogently by Wood Jones[4] in his fine book 'Mammals of South Australia'. He ascribed the decline to many causes, mostly human, but including the displacement of many smaller herbivores by the rabbit and the depredations of the European red fox, *Vulpes vulpes*, and the feral cat. The importance of the fox and cat on marsupials may be exaggerated, for it seems unlikely that these two species could have been so effective, when the marsupials had survived the previous predations of *Thylacinus*, *Sarcophilus* and the dingo. However, all attempts to measure the decline of marsupials, much less to determine their cause, are frustrated by the lack of good information about their abundance at the time of settlement; while the number of species extant at that time is reliable, some were probably already rare and declining to extinction. *Potorous platyops* is an example of such a species, while *Burramys parvus* was only known as a fossil until 1965, when the first living specimen was discovered in the Snowy Mountains.[6] For other species, the accounts of naturalists such as Krefft,[150] give an indication of abundance unknown today. Krefft made a journey down the Murray River in 1852, when he collected 18 species of marsupial in its vicinity; only 5 are likely to be encountered there now, being the grey and red kangaroos, the brush possum, the ringtail possum and *Sminthopsis crassicaudata*.

	Rain forest	Eucalypt forest	Woodland savannah	Grassland plains	Total extinct	Total original species	% extinct
Dasyuridae	0	1	1	3	4	15	27
Peramelidae	0	0	1	2	2	6	33
Phalangeridae and Vombatidae	0	0	1	1	1	12	8
Macropodidae	0	2	1	3	3	19	16
Total extinct	0	3	4	9	10	52	19
Total original species	15	37	31	21			
% extinct	0	8	13	43			
% rare and extinct	27	24	45	86			

Fig. 7.2 Extinction of marsupials in New South Wales, analysed by family and habitat. The disparities in the totals are because some species occurred in more than one habitat. (From Marlow[170])

Present status

The best analysis of the present status of marsupials was made by
Marlow[170] who collected all the records for New South Wales and made
a survey to determine the current status of each species in that state. His
results are analysed according to family and to major habitats in Fig. 7.2
and show that the most severely affected groups are the small insectivorous species and the small macropods, which were common in the
grasslands and savannah woodlands, whereas the phalangerids, which are
forest dwellers, have been affected least. The one species in the latter
group extinct in New South Wales is the plains wombat, *Lasiorhinus
latifrons*, which still lives in South Australia. When the percentage of rare
species is added to those presumed extinct, the disparity between the
forest habitats and the grasslands is even more marked and only 3 of the
grasslands species are likely to be found there today.

The pattern in Western Australia and South Australia is similar and
in these two states, samples of the past fauna are still to be found on some
of the offshore islands, as mentioned in Chapter 3. The account of the
Western Australian Museum expedition to Bernier and Dorre Islands[217]
illustrates the diversity of small marsupials and indigenous rodents and
the abundance of the several species to be found where European influence had not impinged on the habitat.

Calaby[60] has also described a rich mammal fauna in the north eastern
part of New South Wales, which has had over 120 years of European
occupation but where interference has been less drastic than in the drier
parts of the state. Here between the upper Richmond and Clarence rivers
occur thriving populations of 11 species of Macropodidae, 9 of Phalangeridae, 2 of Peramelidae and 6 of Dasyuridae, as well as 15 other
indigenous mammals and 7 introduced species. Further north in Queensland, Finlayson some years before described an equally diverse fauna
and in both areas the important factors in maintaining this state are that
most of the species are adapted to eucalypt forest and that the moun-
tainous topography dissects the country into many different habitats
each optimum for different species. For example, to quote Calaby, 'the
euro, *Macropus robustus*, is found chiefly on stony slopes and hilltops, the
whiptail wallaby, *Macropus parryi*, on slopes and hilltops with sparser
woodland and a thin-stemmed grass understory, and the swamp wallaby,
Wallabia bicolor, in the wetter spots with long dense grass or ferns'. The
human activities in the area are forestry and beef cattle grazing, neither of
which causes long term alteration to the habitat, so that a fruitful
conservation of the wildlife has been achieved.

Effects of forest clearance

A practice, which has a far more drastic effect on the forest-adapted

Plate 12 *(above)* The bandicoot *Perameles bougainville*, from Bernier 1., Western Australia. (H. Tyndale-Biscoe)

Plate 13 *(below)* A murine opossum of the genus *Marmosa* from Cali, Colombia. A pouch is not developed in this genus but the 8 young are each attached firmly to a teat (H. Tyndale-Biscoe, by permission of National Geographic Society)

species is the transformation of wet sclerophyll forest to pine plantation. The dominant species of the native forest in the higher rainfall areas of New South Wales and Queensland are deemed to have a low timber yield and the timber itself is not considered to be of good quality, so it is being removed by clear felling and, after burning off, the country is being planted to high-yielding *Pinus radiata*. During this sequence it is possible to see directly how the indigenous species are affected, and by analogy understand the process that previously overtook the species of the grassland and savannah woodland.

The commonest arboreal species in this forest is the greater glider, *Schoinobates volans*, which was considered in Chapter 4. As part of these studies we also followed the fate of the animals after felling.[282] The area of forest chosen for our study, which lasted 5 years, was almost entirely surrounded by pine plantations and cleared farm land where gliders cannot live, so the marked animals displaced during the first year had only the next year's block in which to find refuge. By marking every animal when caught after tree fall, we were able to determine its short term survival during that felling season and its long term survival if recovered in adjacent blocks during succeeding years. Very few animals were harmed at tree fall because they were able to glide free as the tree went over. Nevertheless, more than three quarters of them were never seen again and of those that were recaptured in the same season, 73% were caught during the next 8 days, when most had lost weight and the females had lost their pouch young. The disappearance could have been due to the others moving well out of the disturbed area but the very few recaptures in subsequent years did not support it. Less than 7% were recovered after 1 year and most of these were of animals originally marked near the boundary of the two blocks. One year an adjacent block of forest was depleted of gliders before felling began with the idea that this would provide unoccupied places for displaced animals to move into, but we did not recover a higher proportion of marked animals from that block than from fully occupied ones. From all of this it seems that when the forest is destroyed over 90% of the resident gliders die on the block; they are not injured physically nor do they move to adjacent forest unless it was part of their previous home range. The fate of other arboreal species is not known but the ground living species do survive. In the same forests that the gliders were studied, J. McIlroy has studied the common wombat, *Vombatus ursinus*. They survive the felling of the forest but in the subsequent years their numbers appear to decline until the forest canopy is re-established by the pine trees, when the population may again increase. Although they subsist on grasses that grow in forest clearings, an abundance of this food is not sufficient to maintain the population in the absence of other factors associated with a closed forest. Brush possums and ringtail possums also become established in pine forests but not so the several species of gliders.

From this consideration of the New South Wales marsupials we may conclude that the survival of the forest dwellers has been due to the fact that this habitat has been the least affected by man, whereas the grassland and savannah woodland has been the most thoroughly altered habitat and its indigenous mammals the most severely affected. Frith and Calaby[1] observe that the intensive stocking of all this country with sheep to a level where the carrying capacity has been declining, from 13 million at the end of the century to 7 million today, must have left little herbage for the indigenous herbivores. As well, the practice of clearing away all cover denies them daytime refuges against heat and predators.

Effects of direct predation

Alteration of habitat has probably been the predominant factor in the decline of marsupials but in a few instances direct human predation may have been a contributory cause. The attempts of farmers to rid their properties of kangaroos and wallabies, have affected disproportionately the species such as the nail-tailed wallaby, *Onychogalea fraenata*, that preferred the open spaces more than species, such as the grey kangaroo, that stay in cover. The decline of the rarer wallabies is clearly seen in Fig. 7.3 by the skins marketed in Queensland during the decade since large scale kangaroo shooting began in that state; the proportion of wallaby skins has declined to a small fraction of the first year, whereas the red kangaroo proportion has remained constant and that of grey kangaroos has increased. Since the total number taken has more than trebled, which represents a greater effort in hunting kangaroos, the figures suggest that the other species have now declined to very low levels.

The thylacine, *Thylacinus cynocephalus*, in Tasmania was declared to be a pest in 1888 and from that year until 1910 a Government bounty of £1 was given for each scalp. The records for the period (Fig. 7.4) show that about 100 were killed each year until 1904 when the numbers quickly fell and none was presented after 1909. However, many more than this were killed for bounties given by local farmers and private companies so that the official total of 2268 is an underestimate of the total number killed up to that time. Guiler[109] considers that the precipitate decline was not caused by the human predation, because this had been exercised at a constant pressure for 20 years, and that disease or habitat alteration must be considered more important. In the absence of all knowledge of the reproduction, fecundity and longevity of thylacines and of their ecology, the primary effect of this concerted human predation cannot be

Year	Number of skins	Macropus giganteus	Megaleia rufa	Macropus robustus	Macropus parryi	Macropus rufogriseus	Wallabia bicolor	Macropus agilis
1955	305 616	51·0	31·6	3·7	3·8	5·1	4·2	1·6
1960	769 948	63·4	32·5	1·3	0·7	1·1	0·9	0·2
1965	1 168 887	67·4	30·2	1·1	0·7	0·4	0·1	0·01

Fig. 7.3   Number of kangaroo and wallaby skins marketed in Queensland, showing the change in percentage of each species. (From Frith and Calaby[1])

so easily discounted. The majority of thylacines was collected in savannah woodland near rocky outcrops and very few from the rain forests of south western Tasmania; since this is the only possible refuge of the species today, it does not augur well for its survival, although occasional recent sightings have been reported.

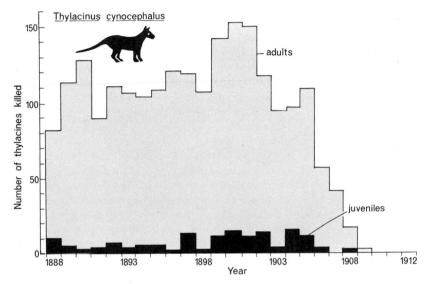

Fig. 7.4   Total of adult and juvenile thylacines presented for Government bounty each year between 1888 and 1912, in Tasmania. (From Guiler[109])

The koala, *Phascolarctos cinereus*, is the best known example of the effects of direct human predation on a marsupial.[6] It was brought to extinction in South Australia by 1920 and to very nearly the same state in Queensland, New South Wales and Victoria. It is now well established again in Victoria but remains rare in the other two States, almost 50 years after its decimation.

Effects of habitat improvement

A few species have increased their range and abundance as a result of human alteration of the habitat. The interaction of the four species of kangaroo with sheep and cattle was mentioned in Chapter 3, where it was noted that the food preferences of the several species only overlap under unfavourable conditions. What was not made clear there was the evidence that pastoral activity favours the red kangaroo and the euro. Frith and Calaby,[1] in reviewing the work of their group in New South

Wales, observe that the extensive grassland community surrounding the Murrumbidgee flood plain is a disclimax maintained by grazing, whereas the original climax association was an open woodland. During aerial counts of this country they observed that 79% of the red kangaroos seen were in the grassland at densities between 3 and 40 times those in the woodland. The land clearance has thus improved the habitat for red kangaroos so that their range has extended further east than in earlier times. Newsome[193] observed analogous results in cattle country near Alice Springs, where the effects of cattle grazing, by inducing new shoots to grow, improves the country for red kangaroos. In the Pilbara of north Western Australia[82] sheep numbers reached a peak of 750 000 in 1930, by which time wells had been placed on all the country; euros and red kangaroos were numerous. Sheep and kangaroos declined in the drought of 1934–5, after which the euro population quickly recovered but the sheep did not. After the next drought in 1944–5 the sheep numbers dropped to 300 000 and both species of kangaroo suffered a heavy mortality. After this drought only the euro recovered and the red kangaroo disappeared from many areas it formerly occupied. By 1955 when Ealey began his study, which had been urged on CSIRO by the pastoral interests, the numbers of euros on all properties far exceeded the number of sheep. For example, Talga Talga station in 1959 carried 2300 sheep but more than 30 000 euros. As we noted in Chapter 3, the euro is able to maintain itself on spinifex, *Triodia*, if water or rock shelters are available, whereas the sheep is not. It is now believed that the stocking levels maintained in the Pilbara until 1930 were greatly in excess of the carrying capacity, with the result that the plant species preferred by the sheep were exterminated and only the ill-favoured species survived; these were precisely the species preferred by euros but not by red kangaroos, so that sheep and red kangaroos could not recover after the droughts but euros could, and did. The abundance of their preferred food plant as well as the artificially increased water supplies provided the means for a great increase in the euro population.

Sheep numbers have likewise declined in New South Wales and in Queensland, possibly for the same reasons, and the red kangaroo is being blamed for it, as the euro was in the Pilbara. If, however, the decline is due to the inability of the pasture to support one species at the past levels, then a mixed grazing by sheep and kangaroos might prove to be more stable. Kangaroos are becoming an important resource for skins as Fig. 7.3 shows; their value as meat producers has so far only been exploited for low grade uses such as pet food. Studies by Tribe and Peel[273] show that the percentage of liveweight composed by muscle protein is 52% in the red kangaroo compared to 32% for cattle and 27% for sheep, so that in this respect their potential value as protein producers is high.

MARSUPIAL COLONIZATION EFFECTED BY MAN

So far in this chapter we have considered the effects of alien species on indigenous marsupials. There are several examples where marsupials are the alien species and have demonstrated the same pattern of invasion as Eutheria have done in Australia.

Didelphis marsupialis in California

The opossum did not penetrate through Baja California during its Pleistocene invasion of North America and was absent from the West Coast until about 1900. People, with a taste for opossum, who immigrated from the eastern states, liberated animals near Los Angeles and in Santa Clara County near San Francisco from 1870 through to 1915. Grinnell, Dixon and Linsdale[107] summarized these early liberations and showed that most of the animals came from Tennessee or Missouri and are thus of the sub-species *virginiana*. By 1932 the opossum was fairly widespread from the southern border of California to a few km north of San Francisco and extended into the foothills (Fig. 7.5). In 1924–5 356 opossums were reported to have been trapped and three years later over 2000 were trapped. By 1958 it had spread[2] north into British Columbia and eastwards everywhere to an altitude of 1500 m. Thus, in 26 years, it had moved at an average rate of 50 km per year, whether by its own powers or by assistance of trappers is not clear. It is interesting that it has not passed the $-7°C$ January isotherm in either Washington State or British Columbia, nor has it joined up with the Mexican population in the south, across the desert regions of Baja California, as these were the two climatic barriers considered in Chapter 6 to be limiting its further spread in the eastern states.

Macropus rufogriseus and other wallabies

The red-necked or Bennett's wallaby is a common animal in Zoos probably because it breeds readily in captivity. It has been liberated several times in Germany during this century and another feral population became established in Derbyshire after escaping from a private Zoo in 1940. In both places the animals thrived for several years and the English population may have reached 500 by 1947.[168] The severe winter then checked the population and the 1962–3 winter reduced the number to 6, from whence they are slowly increasing again. The German populations thrived until the first World War, when they were all shot for food; a similar fate befell a subsequent population established near Hamburg in 1940.[43]

Red necked wallabies were also liberated near Waimate in New

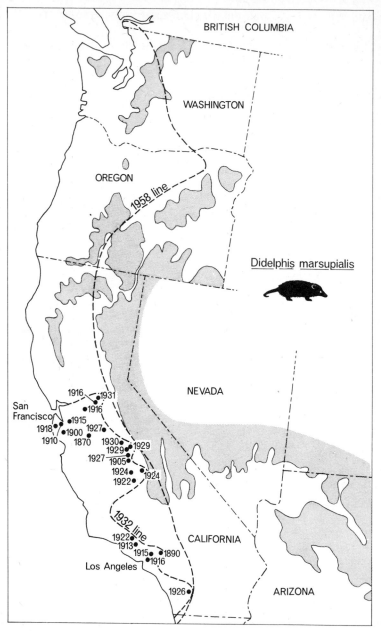

Fig. 7.5 Western north America to show the liberation points of the opossum, *Didelphis marsupialis*, and the extent of spread by 1932 and by 1958. (From Grinnell, Dixon and Linsdale,[107] and Hall and Kelson[2])

Zealand where they are well established, despite continuing efforts to eradicate them.[291] The most notable, if inadvertent liberation of a macropod in New Zealand was made on Kawau Island, Auckland Harbour. Several species were placed on this island in 1870 but their identity was not established until 1967, when it was discovered that one species was *Macropus parma*, considered at the time to have become extinct in Australia. Subsequently one specimen has been collected in New South Wales, but the New Zealand population is now the most abundant.[6]

The only other feral macropods are the rock wallabies, *Petrogale penicillata*, that inhabit the rocky slopes of the Hawaiian Island of Oahu, where two animals escaped in 1916 and a small population is now established in a few valleys.[272]

Trichosurus vulpecula in New Zealand

The brush possum is one of the few marsupials whose range has increased in Australia since European settlement; it has survived continued and fairly heavy trapping for skins, especially from Tasmania. In Chapter 4 we noted that its success in Australia seemed to rest on its adaptability to a variety of foods and habitats. Its qualities as a fur bearer was the motive for its liberation in New Zealand, where no indigenous fur-bearing mammals previously existed. The first liberations were made soon after European settlement began in 1840 and new imports from Tasmania, Victoria and New South Wales continued by private enterprise until 1898, when the Government took an active interest in promoting its establishment.[207] From the start it did extremely well in the New Zealand rain forests as the example given on p. 164 shows.

By natural spread and secondary liberations from New Zealand stock the species had covered much of the area of both islands by 1946 when Wodzicki[291] made the first detailed survey. It was not present north of Auckland, the centre of the north island or in large parts of the south island. Now, 25 years later, it has occupied all suitable habitat and is no longer regarded as a beneficial resource. An attempt to halt its spread and bring it under control in 1950, by offering a bounty for every scalp, was abandoned ten years later when the expenditure was manifestly ineffective. The question of the possum's harmful effects upon the indigenous forests has been debated for decades by loud protagonists on both sides, but with very little detailed evidence. Far more careful study is required to distinguish the effects of the possum from the intrinsic long term changes that were going on in the forests before its introduction. The view that the New Zealand forests have evolved in the absence of browing animals, and are therefore particularly susceptible to animals such as deer and possums, cannot be sustained against the evidence that a variety of moas browsed the forests until 700 years ago and the hosts of leaf eating insects that inhabit New Zealand forests. Holloway's [127] view that the rain forests are a relict of colder, wetter climates and were in process of being replaced by *Nothofagus* forests when the mammals were introduced, has more to commend it. On this hypothesis the mammals are merely accelerating a process already in being.

CONCLUSION

Man's impact on marsupials has been the same as his impact on other species around the world. The large species and the highly specialized succumb and disappear when their particular habitat is altered. A few species thrive under the altered conditions, either by being particularly adaptable or by being pre-adapted to the new habitat that comes to prevail. The disappearance of the marsupials in Australia is often offered as evidence for the inferior status of marsupials vis-a-vis eutherians. Neither in South America, where the marsupials survived man's arrival, nor in Australia where many disappeared, is the pattern of extinction different from the changes in the mammalian fauna of other continents; the real difference is that modern man's impact on the Australian fauna has come much later than in Europe and sufficiently recently for it to have been recorded in some detail.

Appendix

A classification of monotreme and marsupial species mentioned in the text. It is based on the classification proposed by Kirsch[147] and by Ride[6] in which three orders are recognized in the Marsupialia. Extinct species or families are indicated (+).

CLASS MAMMALIA
Sub-class Prototheria

Order Monotremata
 Tachyglossidae
  Tachyglossus aculeatus, echidna
  Zaglossus
 Ornithorhynchidae
  Ornithorhynchus anatinus, platypus

Sub-class Metatheria = Marsupialia

Order Polyprotodonta
 Didelphidae
  + Thlaeodontinae
   + *Alphadon*
   + *Clemensia*
   + *Didelphodon*
   + *Glasbius*
   + *Pediomys*
  Microbiotherinae
   Caluromys derbianus, woolly opossum
   Dromiciops australis
   Glironia venusta, bush-tailed opossum
  Didelphinae
   Chironectes minimus, water opossum
   Didelphis marsupialis, common opossum

Order Polyprotodonta—*cont.*

      D. albiventris (= *azarae, aurita*), white-eared opossum
      Lutreolina crassicaudata, comadrejas
      Marmosa microtarsus, murine opossum
      M. mitis
      M. murina
      Metachirus nudicaudatus, brown 4-eyed opossum
      Monodelphis brevicaudata, short-tailed opossum
      Philander opossum, grey 4-eyed opossum
+ Carolameghinidae
+ Argyrolagidae
+ Polydolopidae
+ Borhyaenidae
    + *Borhyaena*
    + *Thylacosmilus*
Dasyuridae
  Dasyurinae
      Dasyurus viverrinus, native cat
      D. geoffroii, chuditch
      D. hallucatus
      Sarcophilus harrisii, Tasmanian devil
      Phascogale tapoatafa, wambenger
      Dasyuroides byrnei, kowari
      Dasycercus cristicauda, mulgara
      Antechinus flavipes, yellow-footed antechinus
      A. stuartii, brown antechinus
      A. swainsonii
      Sminthopsis crassicaudata, fat-tailed dunnart
      S. froggatti (*larapinta*)
      Antechinomys spenceri, wuhl-wuhl
  Myrmecobiinae
      Myrmecobius fasciatus, numbat or banded anteater
Thylacinidae
      Thylacinus cynocephalus, Tasmanian tiger or thylacine
Notoryctidae
      Notoryctes typhlops, marsupial mole
Peramelidae
  Peramelinae
      Perameles nasuta, long-nosed bandicoot
      P. gunnii, barred bandicoot
      P. bougainville
      Isoodon macrourus, short-nosed bandicoot
      Echimypera
  Macrotinae
      Macrotis lagotis, rabbit-eared bandicoot

Order Paucituberculata
Caenolestidae
      Caenolestes obscurus, rat opossum
      Orolestes inca
      Rhyncholestes raphanurus

Order Diprotodonta
Vombatidae
+ *Phascolonus*
Vombatus ursinus, common wombat
Lasiorhinus latifrons, plains wombat
Phascolarctidae
Phascolarctos cinereus, koala
Wynyardidae
+ *Wynyardia bassiana*
Phalangeridae
Phalanger maculatus, cuscus
P. orientalis
Trichosurus vulpecula, brush possum
T. caninus, mountain possum
Wyulda squamicaudata, scaly-tailed possum
Petauridae
Pseudocheirus peregrinus, ringtail possum
Petaurus breviceps, sugar glider
Schoinobates volans, greater glider
Burramyidae
Acrobates pygmaeus, pigmy glider
Burramys parvus, mountain pigmy possum
Cercartetus concinnus, pigmy possum
C. nanus
C. lepidus
Thylacoleonidae
+ *Thylacoleo carnifex*, marsupial lion
Diprotodontidae
+ *Diprotodon*
+ *Nototherium*
+ *Zygomaturus*
Macropodidae
Potoroinae
Aepyprymnus rufescens, rat kangaroo
Bettongia penicillata, brush-tailed bettong
B. gaimardi
B. lesueur, boodie
Potorous tridactylus, long-nosed potoroo
P. platyops, broad-faced potoroo
Macropodinae
+ *Macropus ferragus*
+ *Propleopus*
Macropus giganteus, eastern grey kangaroo
M. fuliginosus, western grey kangaroo
M. robustus, euro
M. antelopinus, antelope kangaroo
Megaleia rufa, red kangaroo
+ *Protemnodon anak*
Macropus agilis, agile wallaby
M. parryi, pretty-faced wallaby, or whiptail wallaby
M. irma, western brush wallaby
M. eugenii, tammar

Order Diprotodonta—*cont.*

 M. parma, parma wallaby
 M. rufogriseus, red-neck or Bennett's wallaby
 Wallabia bicolor, swamp wallaby
 Thylogale billardierii, pademelon
 T. thetis
 T. stigmatica
 Setonix brachyurus, quokka
 Onychogalea fraenata, nail-tailed wallaby
 Lagorchestes conspicillatus, spectacled hare wallaby
 L. hirsutus, western hare wallaby
 Lagostrophus fasciatus, banded hare wallaby
 Petrogale penicillata, rock wallaby
 P. rothschildi
 Peradorcas concinna
+ Sthenurinae
 + *Sthenurus*, short-faced kangaroo
 + *Procoptodon*
Tarsipedidae
 Tarsipes spencerae, honey possum

References

GENERAL REFERENCES

1. FRITH, H. J. and CALABY, J. H. (1969). *Kangaroos*. Cheshire, Melbourne.
2. HALL, E. R. and KELSON, K. R. (1959). *The Mammals of North America*. **1**, Ronald Press, New York.
3. HARTMAN, C. G. (1952). *Possums*. Univ. Texas Press, Austin.
4. JONES, F. W. (1924). *The Mammals of South Australia*. Government Printer, Adelaide.
5. MARLOW, B. J. (1964). *Marsupials*. Jacaranda, Brisbane.
6. RIDE, W. D. L. (1970). *A Guide to the Native Mammals of Australia*. Oxford Univ. Press, Melbourne.
7. WARING, H., MOIR, R. J. and TYNDALE-BISCOE, C. H. (1966). Comparative physiology of Marsupials. *Adv. Comp. Physiol. Biochem.* **2**, 237-376.

SPECIAL REFERENCES

8. ABBIE, A. A. (1937). Some Observations on the Major Sub-divisions of the Marsupialia. *J. Anat.* **71**, 429–36.
9. ABBIE, A. A. (1939). A Masticatory Adaptation Peculiar to some Diprotodont Marsupials. *Proc. zool. Soc. Lond.* **109**, 261–79.
10. ADAMS, D. M. and BOLLIGER, A. (1954). Observations on carbohydrate metabolism and alloxan diabetes in a marsupial (*Trichosurus vulpecula*). *Aust. J. exp. Biol. med. Sci.* **32**, 101–11.
11. AKERT, K., POTTER, D. and ANDERSON, J. W. (1961). The Subfornical organ in mammals. 1. Comparative and topographical anatomy. *J. comp. Neurol.* **116**, 1–9.
12. AMOROSO, E. C. (1952). *Placentation*. In *Marshall's Physiology of Reproduction*, ed. Parkes, A. S. 3rd edn. Vol. 2. Longmans, Green, London.
13. ANDERSON, D. (1937). Studies on the opossum (*Trichosurus vulpecula*). 1. Blood analysis and lipoid glandular constituents in normal and lactating opossums. 2. The effects of splenectomy, adrenalectomy and injections of cortical hormones. *Aust. J. exp. Biol. med. Sci.* **15**, 17–32.

14. ANDERSON, D. (1938). Diastase in *Trichosurus*. *Aust. J. exp. Biol. med. Sci.* **16**, 117–32.
15. ARNOLD, J. and SHIELD, J. (1970). Oxygen consumption and body temperature of the Chuditch (*Dasyurus geoffroii*). *J. Zool. Lond.* **160**, 391–404.
16. BAILEY, L. F. and LEMON, M. (1966). Specific milk proteins associated with resumption of development by the quiescent blastocyst of the lactating Red Kangaroo. *J. Reprod. Fert.* **11**, 473–75.
17. BAKER, E. and SIMMONDS, W. J. (1966). Membrane ATPase and electrolyte levels in marsupial erythrocytes. *Biochim. biophys. Acta.* **126**, 492–99.
18. BARKER, J. M. (1961a). The metabolism of carbohydrate and volatile fatty acids in the marsupial, *Setonix brachyurus*. *Quart. Jl. exp. Physiol.* **46**, 54–68.
19. BARKER, J. M. (1961b). Properties of the erythrocytes of two marsupials: the brush-tail possum, *Trichosurus vulpecula* (Kerr) and the quokka, *Setonix brachyurus* (Quoy and Gaimard). *Quart. Jl. exp. Physiol.* **46**, 123–30.
20. BARKER, S. (1961a). Copper, molybdenum and inorganic sulphate levels in Rottnest plants. *J. Proc. R. Soc. West Aust.* **44**, 49–52.
21. BARKER, S. (1961b). Studies on marsupial nutrition. 3. The copper-molybdenum-inorganic sulphate interaction in the Rottnest Quokka, *Setonix brachyurus* (Quoy and Gaimard). *Aust. J. biol. Sci.* **14**, 646–58.
22. BARKER, S. (1962). Copper-levels in the milk of a marsupial. *Nature, Lond.* **193**, 292.
23. BARKER, S. (1968). Nitrogen balance and water intake in the Kangaroo Island wallaby, *Protemnodon eugenii* (Desmarest). *Aust. J. exp. Biol. med. Sci.* **46**, 17–32.
23A. BARKER, S. (1971). Nitrogen and water excretion of wallabies: differences between field and laboratory findings. *Comp. Biochem. Physiol.* **38A**, 359–67.
24. BARKER, S., BROWN, G. D. and CALABY, J. H. (1963). Food Regurgitation in the Macropodidae. *Aust. J. Sci.* **25**, 430–32.
25. BARKER, S., LINTERN, S. M. and MURPHY, C. R. (1970). The effect of water restriction on urea retention and nitrogen excretion in the Kangaroo Island wallaby, *Protemnodon eugenii* (Desmarest). *Comp. Biochem. Physiol.* **34**, 883–93.
26. BARNARD, E. A. (1969). Biological function of Pancreatic Ribonuclease. *Nature, Lond.* **221**, 340–44.
27. BARNETT, C. H. and BRAZENOR, C. W. (1958). The testicular rete mirabile of marsupials. *Aust. J. Zool.* **6**, 27–32.
28. BARTHOLOMEW, G. A. (1956). Temperature regulation in the macropod marsupial, *Setonix brachyurus*. *Physiol. Zool.* **29**, 26–40.
29. BARTHOLOMEW, G. A. and HUDSON, J. W. (1962). Hibernation, estivation, temperature regulation, evaporative water loss, and heart rate of the pigmy possum, *Cercartetus nanus*. *Physiol. Zool.* **35**, 94–107.
30. BASEDOW, H. (1914). Aboriginal rock carvings of great antiquity in South Australia. *J. Roy. Anthrop. Inst.* **44**, 195–211.
31. BAUMAN, T. R. and TURNER, C. W. (1966). L–Thyroxine secretion rates and L–triiodothyronine equivalents in the opossum, *Didelphis virginianus*. *Gen. comp. Endocr.* **6**, 109–13.

32. BAUTISTA, N. S. and MATZKE, H. A. (1965). A degeneration study of the course and extent of the pyramidal tract of the opossum. *J. comp. Neurol.* **124**, 367–76.
33. BENTLEY, P. J. (1955). Some aspects of the water metabolism of an Australian marsupial *Setonyx brachyurus*. *J. Physiol., Lond.* **127**, 1–10.
34. BENTLEY, P. J. (1960). Evaporative water loss and temperature regulation in the marsupial *Setonyx brachyurus*. *Aust. J. exp. Biol. med. Sci.* **38**, 301–6.
35. BENTLEY, P. J. and SHIELD, J. W. (1962). Metabolism and Kidney Function in the Pouch Young of the Macropod Marsupial *Setonix brachyurus*. *J. Physiol., Lond.* **164**, 127–37.
36. BERGMAN, H. C. and HOUSLEY, C. (1968). Chemical analyses of American opossum (*Didelphys virginiana*) milk. *Comp. Biochem. Physiol.* **25**, 213–18.
37. BERKOVITZ, B. K. B. (1966). The homology of the premolar teeth in *Setonix brachyurus* (Macropodidae: Marsupialia). *Archs. oral Biol.* **11**, 1371–84.
38. BIGGERS, J. D. (1966). Reproduction in male marsupials. *Symp. zool. Soc. Lond.* **15**, 251–80.
39. BLAIR-WEST, J. R., COGLAN, J. P., DENTON, D. A., NELSON, J. F., ORCHARD, E., SCOGGINS, B. A., WRIGHT, R. D., MYERS, K. and JUNQUEIRA, C. L. (1968). Physiological, morphological and behavioural adaptation to a sodium deficient environment by wild native Australian and introduced species of animals. *Nature, Lond.* **217**, 922–28.
40. BLOCK, M. (1960). Wound healing in the New-Born Opossum (*Didelphis virginianam*). *Nature, Lond.* **187**, 340–1.
41. BLOCK, M. (1964). The Blood Forming Tissues and Blood of the New-Born Opossum (*Didelphys virginiana*). 1. Normal Development Through About the One Hundreth Day of Life. *Ergebn. Anat. Entw-Gesch.* **37**, 237–366.
42. BODIAN, D. (1939). Studies on the diencephalon of the Virginia opossum. *J. comp. Neurol.* **71**, 259–323.
43. BOETTGER, C. R. (1943). Kanguruhs in Deutschland in freier Wildbahn. *Natur. Volk.* **73**, 331–36.
44. BOLLIGER, A. and HARDY, M. H. (1944). The Sternal Integument of *Trichosurus vulpecula*. *J. Proc. R. Soc. N.S.W.* **78**, 122–33.
45. BOLLIGER, A. and SHORLAND, F. B. (1963). Investigation on Fats of Australian Mammals. *Aust. J. Sci.* **25**, 453–66.
46. BORCH, S. M. VON DER. (1963). Unilateral hormone effect in the marsupial *Trichosurus vulpecula*. *J. Reprod. Fert.* **5**, 447–49.
47. BOURNE, G. H. (1934). Unique structure in the adrenal of the female opossum. *Nature, Lond.* **134**, 664–65.
48. BRAMBELL, F. W. R. (1958). The Passive Immunity of the Young Mammal. *Biol. Rev.* **33**, 488–531.
49. BRAMBELL, F. W. R. and HALL, K. (1937). Reproduction in the lesser shrew (*Sorex minutus* Linneaus). *Proc. zool. Soc. Lond.* **1936**, 957–69.
50. BRITTON, S. W. and SILVETTE, H. (1937). Further observations on sodium chloride balance in the adrenalectomised opossum. *Am. J. Physiol.* **118**, 21–25.
51. BROCKE, R. H. (1970). Ecological inferences from oxygen consumption data of the opossum. *Bull. ecol. Soc. Am.* **51**, 29.

52. BROOM, R. (1900). Development and morphology of the marsupial shoulder girdle. *Trans. R. Soc. Edinb.* **39**, 749–70.

53. BROWN, G. D. (1969). Studies on marsupial nutrition. 6. The utilization of dietary urea by the euro or hill kangaroo, *Macropus robustus* (Gould). *Aust. J. Zool.* **17**, 187–94.

54. BROWN, G. D. and MAIN, A. R. (1967). Studies on marsupial nutrition. 5. The nitrogen requirements of the euro, *Macropus robustus. Aust. J. Zool.* **15**, 7–27.

55. BUCHANAN, G. D. (1969). Reproduction in the ferret (*Mustela furo*). 2. Changes following ovariectomy during early pregnancy. *J. Reprod. Fert.* **18**, 305–16.

56. BURKE, J. D. (1954). Blood volume in mammals. *Physiol. Zool.* **27**, 1–21.

57. BUTTLE, J. M., KIRK, R. L. and WARING, H. (1952). Effect of complete adrenalectomy on the wallaby (*Setonix brachyurus*). *J. Endocr.* **8**, 281–90.

58. CSIRO FILM UNIT. (1965). *Birth of the Red Kangaroo.* C.S.I.R.O., Melbourne.

59. CALABY, J. H. (1958). Studies in marsupial nutrition. 2. The rate of passage of food residues and digestibility of crude fibre and protein by the quokka, *Setonix brachyurus* (Quoy and Gaimard). *Aust. J. biol. Sci.* **11**, 571–80.

60. CALABY, J. H. (1966). Mammals of the Upper Richmond and Clarence Rivers, New South Wales. *Tech. Pap. Div. Wildl. Surv. C.S.I.R.O. Aust.* No. 10. 1–55.

61. CAUGHLEY, G. J. (1964). Density and dispersion of two species of kangaroo in relation to habitat. *Aust. J. Zool.* **12**, 238–49.

62. CHENG, C. C. (1955). The development of the shoulder region of the opossum, *Didelphys virginiana*, with special reference to the musculature. *J. Morph.* **97**, 415–71.

63. CLARK, M. J. (1967). Pregnancy in the lactating Pigmy Possum *Cercartetus concinnus. Aust. J. Zool.* **15**, 673–83.

64. CLARK, M. J. and POOLE, W. E. (1967). The Reproductive System and Embryonic Diapause in the Female Grey Kangaroo, *Macropus giganteus. Aust. J. Zool.* **15**, 441–59.

65. CLEMENS, W. A. (1968). Origin and early evolution of marsupials. *Evolution N.Y.* **22**, 1–18.

66. COGHLAN, J. P. and SCOGGINS, B. A. (1967). The measurement of aldosterone, cortisol and corticosterone in the blood of the wombat (*Vombatus hirsutus* Perry) and the kangaroo (*Macropus giganteus*). *J. Endocr.* **39**, 445–48.

67. COOK, B. and NALBANDOV, A. V. (1968). The effect of some pituitary hormones on progesterone synthesis *in vitro* by the luteinized ovary of the common opossum (*Didelphis marsupialis virginiana*). *J. Reprod. Fert.* **15**, 267–75.

68. COX, C. B. (1970). Migrating marsupials and drifting continents. *Nature, Lond.* **226**, 767–70.

69. CROWCROFT, P. and GODFREY, G. K. (1968). The daily cycle of activity in two species of *Sminthopsis* (Marsupialia: Dasyuridae). *J. Anim. Ecol.* **37**, 63–73.

70. DAWSON, T. J. (1969). Temperature regulation and evaporative water loss in the brush-tailed possum *Trichosurus vulpecula. Comp. Biochem. Physiol.* **28**, 401–7.

71. DAWSON, T. J. and BROWN, G. D. (1970). A comparison of the insulative and reflective properties of the fur of desert kangaroos. *Comp. Biochem. Physiol.* **37**, 23–38.

72. DAWSON, T. J. and DENNY, M. J. S. (1968). The influence of the spleen on blood volume and haematocrit in the brush-tailed possum (*Trichosurus vulpecula*). *Aust. J. Zool.* **16**, 603–8.

73. DAWSON, T. J. and DENNY, M. J. S. (1969a). A bioclimatological comparison of the summer day microenvironments of two species of Arid Zone kangaroo. *Ecology*, **50**, 328–32.

74. DAWSON, T. J. and DENNY, M. J. S. (1969b). Seasonal variation in the plasma and urine electrolyte concentration of the arid zone kangaroos *Megaleia rufa* and *Macropus robustus*. *Aust. J. Zool.* **17**, 777–84.

75. DAWSON, T. J. and HULBERT, A. J. (1970). Standard metabolism, body temperature, and surface areas of Australian marsupials. *Am. J. Physiol.* **218**, 1233–38.

76. DAWSON, T. J., DENNY, M. J. S. and HULBERT, A. J. (1969). Thermal balance of the macropodid marsupial *Macropus eugenii* Desmarest. *Comp. Biochem. Physiol.* **31**, 645–53.

77. DIETZ, R. S. and HOLDEN, J. C. (1970). The break up of Pangaea. *Scient. Am.* **223**, 30–41.

78. DOLPH, C. I., BRAUN, H. A. and PFEIFFER, E. W. (1962). The effect of vasopressin upon urine concentration in *Aplodontia rufa* (Sewellel) and the rabbit. *Physiol. Zool.* **35**, 263–69.

79. DUNNET, G. M. (1962). A population study of the quokka, *Setonix brachyurus* (Quoy and Gaimard) (Marsupialia). 2. Habitat, movements, breeding and growth. *C.S.I.R.O. Wildl. Res.* **7**, 13–32.

80. DUNNET, G. M. (1964). A field study of local populations of the brush-tailed possum *Trichosurus vulpecula* in Eastern Australia. *Proc. zool. Soc. Lond.* **142**, 665–95.

81. EALEY, E. H. M. (1963). The ecological significance of delayed implantation in a population of the hill kangaroo (*Macropus robustus*). In *Delayed Implantation* ed. Enders, A. C. University Press, Chicago.

82. EALEY, E. H. M. (1967). Ecology of the Euro, *Macropus robustus* (Gould) in north-western Australia. *C.S.I.R.O. Wildl. Res.* **12**, 9–51.

83. EALEY, E. H. M., BENTLEY, P. J. and MAIN, A. R. (1965). Studies on water metabolism of the Hill kangaroo, *Macropus robustus* (Gould), in North-west Australia. *Ecology* **46**, 473–79.

84. EBNER, F. F. (1967). Afferent connections to neocortex in the opossum (*Didelphis virginiana*). *J. comp. Neurol.* **129**, 241–68.

85. ENDERS, A. C. (Ed.) (1963). *Delayed Implantation*. University Press, Chicago.

86. ENDERS, A. C. and ENDERS, R. K. (1969). The placenta of the four-eyed possum (*Philander opossum*). *Anat. Rec.* **165**, 431–50.

87. ENDERS, R. K. (1966). Attachment, nursing and survival of young in some didelphids. *Symp. zool. Soc. Lond.* **15**, 195–203.

88. EZEKIEL, E. (1963). Exchange *in vitro* between specific iron-binding proteins of foetal and maternal blood plasma and milk whey in the albino rat and a marsupial. *Biochim. biophys. Acta.* **78**, 223–25.

89. FLYNN, T. T. (1923). The Yolk-Sac and Allantoic Placenta in *Perameles*. *Q. Jl. microsc. Sci.* **67**, 123–82.

90. FLYNN, T. T. (1930). The uterine cycle of pregnancy and pseudo-pregnancy as it is in the diprotodont marsupial *Bettongia cuniculus*, with notes on other reproductive phenomena in this marsupial. *Proc. Linn. Soc. N.S.W.* **55**, 506–31.

91. FOOT, J. Z. and ROMBERG, B. (1965). The utilization of roughage by sheep and the red kangaroo, *Macropus rufus* (Desmarest). *Aust. J. agric. Res.* **16**, 429–35.

92. FORBES, D. K. and TRIBE, D. E. (1969). Salivary glands of Kangaroos. *Aust. J. Zool.* **17**, 765–75.

93. FORBES, D. K. and TRIBE, D. E. (1970). The utilization of roughages by sheep and kangaroos. *Aust. J. Zool.* **18**, 247–56.

94. FRANCQ, E. N. (1969). Behavioural aspects of feigned death in the opossum *Didelphis marsupialis*. *Am. Midl. Nat.* **81**, 556–68.

95. FRASER, E. H. and KINNEAR, J. E. (1969). Urinary creatinine excretion by macropod marsupials. *Comp. Biochem. Physiol.* **28**, 685–92.

96. FRITH, H. J. and SHARMAN, G. B. (1964). Breeding in Wild populations of the Red Kangaroo, *Megaleia rufa*. *C.S.I.R.O. Wildl. Res.* **9**, 86–114.

97. GERSH, I. (1937). The correlation of structure and function in the developing mesonephros and metanephros. *Contr. Embryol. Carneg. Inst.* **26**, 35–58.

98. GILMORE, D. P. (1969). Seasonal reproductive periodicity in the male Australian brush-tailed possum (*Trichosurus vulpecula*). *J. Zool. Lond.* **157**, 75–98.

99. GILMORE, D. P. (1970). The rate of passage of food in the brush-tailed possum, *Trichosurus vulpecula*. *Aust. J. biol. Sci.* **23**, 515–18.

100. GODFREY, G. K. (1968). Body-temperatures and torpor in *Sminthopsis crassicaudata* and *S. larapinta* (Marsupialia: Dasyuridae). *J. Zool. Lond.* **156**, 499–511.

101. GODFREY, G. K. (1969). Reproduction in a laboratory colony of the marsupial mouse *Sminthopsis larapinta* (Marsupialia: Dasyuridae). *Aust. J. Zool.* **17**, 637–54.

102. GRIFFITHS, M. (1968). *Echidnas*. Pergamon Press, Oxford.

103. GRIFFITHS, M. and BARKER, R. (1966). The plants eaten by sheep and by kangaroos grazing together in a paddock in south-western Queensland. *C.S.I.R.O. Wildl. Res.* **11**, 145–67.

104. GRIFFITHS, M. and BARTON, A. A. (1966). The ontogeny of the stomach in pouch young of the red kangaroo. *C.S.I.R.O. Wildl. Res.* **11**, 169–85.

105. GRIFFITHS, M., MCINTOSH, D. L. and LECKIE, R. M. C. (1969). The effects of cortisone on nitrogen balance and glucose metabolism in diabetic and normal kangaroos, sheep and rabbits. *J. Endocr.* **44**, 1–12.

106. GRIFFTHS, M., MCINTOSH, D. L. and LECKIE, R. M. C. (1972). The mammary glands of the red kangaroo, with observations on the fatty acid components of the milk triglycerides. *J. Zool. Lond.* **166**, 265–75.

107. GRINNELL, J., DIXON, J. S. and LINSDALE, J. M. (1937). *Fur-bearing mammals of California*. Vol. 1. Univ. California, Berkeley.

108. GROSS, R. and BOLLIGER, A. (1959). Composition of milk of the mar-supial *Trichosurus vulpecula*. *Am. J. Dis. Child.* **98**, 768–75.

109. GUILER, E. R. (1961). The former distribution and decline of the Thylacine. *Aust. J. Sci.* **23**, 207–10.

110. GUILER, E. R. and HEDDLE, R. W. L. (1970). Testicular and body temperatures in the Tasmanian devil and three other species of marsupial. *Comp. Biochem. Physiol.* **33**, 881–91.

111. GUNSON, M. M., SHARMAN, G. B. and THOMSON, J. A. (1968). The affinities of *Burramys* (Marsupialia: Phalangeroidea) as revealed by a study of its chromosomes. *Aust. J. Sci.* **31**, 40–41.

112. HAMMOND, J. and MARSHALL, F. H. A. (1952). *The Life Cycle.* In *Marshall's Physiology of Reproduction.* ed. Parkes, A. S. 3rd edn. Vol. 2. Longmans, Green, London.

113. HARRISON, R. G. (1949). The comparative anatomy of the blood-supply of the mammalian testis. *Proc. zool. Soc. Lond.* **119**, 325–42.

114. HARTMAN, C. (1923). The oestrous cycle in the opossum. *Am. J. Anat.* **32**, 353–421.

115. HARTMAN, C. G. (1925). The interruption of pregnancy by ovariectomy in the aplacental opossum: a study in the physiology of implantation. *Am. J. Physiol.* **71**, 436–54.

116. HAYMAN, D. L. and MARTIN, P. G. (1965). An autoradiographic study of D.N.A. synthesis in the Sex Chromosomes of two Marsupials with an XX/XY$_1$Y$_2$ Sex Chromosome mechanism. *Cytogenetics* **4**, 209–18.

116A. HEARN, J. P. (1972). The development of a radioimmunoassay for gonadotrophin in the tammar wallaby, *Macropus eugenii. J. Reprod. Fert.,* **28**, 132.

117. HERSHKOVITZ, P. (1969). The evolution of mammals on southern continents. 6. The recent mammals of the neotropical region: a zoogeographic and ecological review. *Q. Rev. Biol.* **44**, 1–70.

118. HICKMAN, V. V. and HICKMAN, J. L. (1960). Notes on the habits of the Tasmanian doormouse phalangers *Cercartetus nanus* (Desmarest) and *Eudromicia lepida* (Thomas). *Proc. zool. Soc. Lond.* **135**, 365–74.

119. HIGGINBOTHAM, A. C. and KOON, W. E. (1955). Temperature regulation in the Virginia opossum. *Am. J. Physiol.* **181**, 69–71.

120. HILL, J. P. (1900a). On the foetal membranes, placentation and parturition of the native cat (*Dasyurus viverrinus*). *Anat. Anz.* **18**, 364–73.

121. HILL, J. P. (1900b). Contributions to the morphology and development of the female urogenital organs in the Marsupialia. *Proc. Linn. Soc. N.S.W.* **25**, 519–32.

122. HILL, J. P. (1910). Contributions to the embryology of the Marsupialia: 4. The early development of the Marsupialia with special reference to the native cat (*Dasyurus viverrinus*). *Q. Jl. microsc. Sci.* **56**, 1–134.

123. HILL, J. P. (1918). Some observations on the early development of *Didelphys aurita. Q. Jl. microsc. Sci.* **63**, 91–139.

124. HILL, J. P. and HILL, W. C. O. (1955). The growth stages of the pouch young of the native cat (*Dasyurus viverrinus*), together with observations on the anatomy of the new-born young. *Trans. zool. Soc. Lond.* **28**, 349–453.

125. HILL, J. P. and O'DONOGHUE, C. H. (1913). The Reproductive Cycle in the Marsupial *Dasyurus viverrinus. Q. Jl. microsc. Sci.* **59**, 133–74.

126. HINKS, N. T. and BOLLIGER, A. (1957). Glucuronuria in a herbivorous marsupial *Trichosurus vulpecula. Aust. J. exp. Biol. med. Sci.* **35**, 37–44.

127. HOLLOWAY, J. (1954). Forests and climates in the south island of New Zealand. *Trans. R. Soc. N.Z.* **82**, 329–410.

128. HOLSWORTH, W. N. (1967). Population dynamics of the quokka, *Setonix brachyurus*, on the west end of Rottnest Island, Western Australia. 1. Habitat and distribution of the quokka. *Aust. J. Zool.* **15**, 29–46.

129. HONIGMANN, H. (1937). Studies of nutrition of mammals. 10. Experiments with Australian silver-grey opossums. *Proc. zool. Soc. Lond.* **111**, 1–35.

130. HOPE, R. M. and FINNEGAN, D. J. (1970). A serum amylase polymorphism in populations of the brush-tailed possum *Trichosurus vulpecula*. *Aust. J. biol. Sci.* **23**, 235–39.

131. HOWARTH, V. S. (1950). Experimental Prostatectomy in a Marsupial (*Trichosurus vulpecula*). *Med. J. Aust.* **2**, 325–30.

132. HUDSON, J. W. and BARTHOLOMEW, G. A. (1964). *Terrestrial animals in dry heat: estivators*. In *Handbook of Physiology* Vol. 5. Chapter 34. Am. Physiol. Soc.

133. HUGHES, R. D. (1965). On the Age Composition of a Small Sample of Individuals from a Population of the Banded Hare Wallaby, *Lagostrophus fasciatus* (Peron and Lesueur). *Aust. J. Zool.* **13**, 75–95.

134. HUGHES, R. L. (1962). Role of the corpus luteum in marsupial reproduction. *Nature, Lond.* **194**, 890–91.

135. HUGHES, R. L. (1965). Comparative Morphology of Spermatozoa from five Marsupial Families. *Aust. J. Zool.* **13**, 533–43.

136. HUGHES, R. L., THOMSON, J. A. and OWEN, W. H. (1965). Reproduction in Natural Populations of the Australian Ringtail Possum, *Pseudocheirus peregrinus* (Marsupialia: Phalangeridae) in Victoria. *Aust. J. Zool.* **13**, 383–406.

137. JOHNSTON, C. I., DAVIS, J. O. and HARTROFT, P. M. (1967). Renin-angiotensin system, adrenal steroids and sodium depletion in a primitive mammal, the American opossum. *Endrocrinology* **81**, 633–42.

138. JONES, I. C., VINSON, G. P., JARRETT, I. G. and SHARMAN, G. B. (1964). Steriod components in the adrenal venous blood of *Trichosurus vulpecula* (Kerr). *J. Endocr.* **30**, 149–50.

139. JORDAN, S. M. and MORGAN, E. H. (1968). The changes in the serum and milk whey proteins during lactation and suckling in the quokka (*Setonix brachyurus*). *Comp. Biochem. Physiol.* **25**, 271–83.

140. KALDOR, I. and EZEKIEL, E. (1962). Iron content of mammalian breast milk: measurements in the rat and in a marsupial. *Nature, Lond.* **196**, 175.

141. KALDOR, I. and EZEKIEL, E. (1964). Milk iron and plasma iron transport in a lactating marsupial during short-term intravenous infusion of iron. *Aust. J. exp. Biol. med. Sci.* **42**, 54–61.

142. KEAST, A. (1968). Evolution of mammals on Southern continents. 4. Australian mammals: Zoogeography and evolution. *Q. Rev. Biol.* **43**, 373–408.

143. KERRY, K. R. (1969). Intestinal disacchararidase activity in a monotreme and eight species of marsupials (with an added note on the disaccharidases of five species of sea birds). *Comp. Biochem. Physiol.* **29**, 1015–22.

144. KINGSMILL, E. (1962). An investigation of criteria for estimating age in the marsupials *Trichosurus vulpecula* Kerr and *Perameles nasuta* Geoffroy. *Aust. J. Zool.* **10**, 597–616.

145. KINNEAR, J. E. and BROWN, G. D. (1967). Minimum heart rates of marsupials. *Nature, Lond.* **215**, 1501.

146. KINNEAR, J. E., PUROHIT, K. G. and MAIN, A. R. (1968). The ability of the tammar wallaby (*Macropus eugenii*, Marsupialia) to drink sea water. *Comp. Biochem. Physiol.* **25**, 761–82.

147. KIRSCH, J. A. W. (1968). Prodromus of the Comparative serology of Marsupialia. *Nature, Lond.* **217**, 418–20.

148. KIRSCH, J. A. W. and POOLE, W. E. (1967). Serological Evidence for speciation in the Grey Kangaroo, *Macropus giganteus*, Shaw 1790 (Marsupialia: Macropodidae). *Nature, Lond.* **215**, 1097.

149. KLEIBER, M. (1961). *The fire of life: An introduction to Animal Energetics.* Wiley, New York.

150. KREFFT, G. (1866). On the vertebrated animals of the lower Murray and Darling, their habits, economy and geographical distribution. *Trans. Phil. Soc. N.S.W.* **1862–1865**, 1–33.

151. LANGWORTHY, O. R. (1925). The development of progression and posture in young opossums. *Am. J. Physiol.* **74**, 1–13.

152. LAPLANTE, E. S., BURRELL, R., WATNE, A. L., TAYLOR, D. L. and ZIMMERMANN, B. (1969). Skin allograft studies in the pouch young of the opossum. *Transplantation* **7**, 67–72.

153. LA VIA, M. F., ROWLANDS, D. T. and BLOCK, M. (1963). Antibody formation in embryos. *Science, N.Y.* **140**, 1219–20.

154. LAY, D. W. (1942). Ecology of the opossum in eastern Texas. *J. Mammal.* **23**, 147–59.

155. LEMON, M. and BARKER, S. (1967). Changes in milk composition of the Red Kangaroo, *Megaleia rufa* (Desmarest), during lactation. *Aust. J. exp. Biol. med. sci.* **45**, 213–19.

156. LEMON, M. and POOLE, W. E. (1969). Specific proteins in the whey from milk of the grey kangaroo. *Aust. J. exp. Biol. med. Sci.* **47**, 283–85.

157. LENDE, R. A. (1963a). Sensory representation in the cerebral cortex of the opossum (*Didelphis virginiana*). *J. comp. Neurol.* **121**, 395–403.

158. LENDE, R. A. (1963b). Motor representation in the cerebral cortex of the opossum (*Didelphis virginiana*). *J. comp. Neurol.* **121**, 405–15.

159. LILLEGRAVEN, J. A. (1969). Latest Cretaceous Mammals of upper part of Edmonton formation of Alberta, Canada, and review of Marsupial-placental dichotomy in mammalian evolution. *Paleont. Contr. Univ. Kans.* **50**, 1–122.

160. LINTERN, S. M. and BARKER, S. (1969). Renal retention of urea in the Kangaroo Island wallaby, *Protemnodon eugenii* (Desmarest). *Aust. J. exp. Biol. med. Sci.* **47**, 243–50.

161. LLEWELLYN, L. M. and DALE, F. H. (1964). Notes on the ecology of the opossum in Maryland. *J. Mammal.* **45**, 113–22.

162. LOO, Y. T. (1930). The forebrain of the opossum, *Didelphis virginiana*. *J. comp. Neurol.* **51**, 13–64.

163. LYNE, A. G. (1967). *Marsupials and Monotremes of Australia.* Angus and Robertson, Sydney.

164. MCCRADY, E. (1938). The embryology of the opossum. *Am. anat. Mem.* **16**, 1–233.

165. MCINTOSH, D. L. (1966). The digestibility of two roughages and the rates of passage of their residues by the red kangaroo, *Megaleia rufa* (Desmarest), and the merino sheep. *C.S.I.R.O. Wildl. Res.* **11**, 125–35.

166. MACLEAN, L. (1967). A note on the longevity and territoriality of *Trichosurus vulpecula* (Kerr) in the wild. *C.S.I.R.O. Wildl. Res.* **12**, 81–82.

166A. MCMANUS, J. J. (1969). Temperature regulation in the opossum, *Didelphis marsupialis virginiana*. *J. Mammal.* **50**, 550–58.

167. MAIN, A. R. (1961). The occurrence of Macropodidae on Islands and its climatic and ecological implications. *J. Proc. R. Soc. West. Aust.* **44**, 84–89.

168. MALLON, D. (1970). Britain's wild wallabies. *Animals* **13**, 256.

169. MALVIN, R. L., WILDE, W. S. and SULLIVAN, L. P. (1958). Localisation of nephron transport by stop flow analysis. *Am. J. Physiol.* **194**, 135–42.

170. MARLOW, B. J. (1958). A survey of the Marsupials of New South Wales. *C.S.R.I.O. Wildl. Res.* **3**, 71–114.

171. MARLOW, B. J. (1961). Reproductive behaviour of the marsupial mouse, *Antechinus flavipes* (Waterhouse) (Marsupialia), and the development of the pouch young. *Aust. J. Zool.* **9**, 203–18.

172. MARTAN, J. and ALLEN, J. M. (1965). The Cytological and Chemical Organisation of the Prostatic Epithelium of *Didelphis virginiana* Kerr. *J. exp. Zool.* **159**, 209–30.

173. MARTIN, C. J. (1903). Thermal adjustment and respiratory exchange in monotremes and marsupials—A study in the development of homoeothermism. *Phil. Trans. R. Soc. Ser. B.* **195**, 1–37.

174. MARTIN, G. F. and KING, J. S. (1968). The basilar pontine gray of the opossum (*Didelphis virginiana*). *J. comp. Neurol.* **133**, 447–62.

175. MARTIN, P. G. and HAYMAN, D. L. (1966). A Complex Sex-chromosome system in the Hare-wallaby *Lagorchestes conspicillatus* Gould. *Chromosoma* **19**, 159–75.

176. MARTIN, P. G. and HAYMAN, D. L. (1967). Quantitative comparisons between the Karyotypes of Australian Marsupials from three different super-families. *Chromosoma* **20**, 290–310.

177. MARTIN, P. S. and WRIGHT, H. E. eds. (1967). *Pleistocene Extinctions— The Search for a Cause*. Yale Univ. Press, Newhaven.

178. MERCHANT, J. C. and SHARMAN, G. B. (1966). Observations on the attachment of Marsupial Pouch Young to the teats and on the rearing of pouch young by foster-mothers of the same or different species. *Aust. J. Zool.* **14**, 593–609.

179. MERRILEES, D. (1968). Man the destroyer; late Quaternary changes in the Australian Marsupial Fauna. *J. Proc. R. Soc. West Aust.* **51**, 1–24.

180. MILLER, J. F. A. P., BLOCK, M., ROWLANDS, D. T. and KIND, P. (1965). Effect of thymectomy on hemopoietic organs of the opossum "embryo". *Proc. Soc. exp. Biol. Med.* **118**, 916–21.

181. MINCHIN, A. K. (1937), Notes on the weaning of a young Koala (*Phascolarctus cinereus*). *Rec. S. Aust. Mus.* **6**, 1–3.

182. MITCHELL, H. H. (1962). *Comparative nutrition of man and domestic animals*. vol. 1. Academic Press, New York and London.

183. MOIR, R. J. (1968). *Ruminant digestion and evolution*. In *Handbook of Physiology*. Section 6. Alimentary Canal. vol. 5. Am. Physiol. Soc.

184. MOIR, R. J., SOMERS, M. and WARING, H. (1956). Studies on marsupial nutrition. 1. Ruminant-like digestion in a herbivorous marsupial *Setonix brachyurus* (Quoy and Gaimard). *Aust. J. biol. Sci.* **9**, 293–304.

185. MORRISON, P. and MCNAB, B. K. (1962), Daily torpor in a Brazilian murine opossum (*Marmosa*). *Comp. Biochem. Physiol.* **6**, 57–68.

186. MORRISON, P. R. (1946). Temperature regulation in three central American mammals. *J. cell. comp. Physiol.* **27**, 125–37.
187. MORRISON, P. R. (1965). Body temperatures in some Australian mammals. 4. Dasyuridae. *Aust. J. Zool.* **13**, 173–87.
188. MULVANEY, D. J. (1969). *Prehistory of Australia.* Thames and Hudson, London.
189. NARDONE, R. M., WILBER, C. G. and MUSACCHIA, X. J. (1955). Electrocardiogram of opossum during exposure to cold. *Am. J. Physiol.* **181**, 352–56.
190. NELSON, L. R. and LENDE, R. A. (1965). Interhemispheric responses in the opossum. *J. Neurophysiol.* **28**, 189–99.
191. NEWSOME, A. E. (1964). Anoestrus in the red kangaroo *Megaleia rufa* (Desmarest). *Aust. J. Zool.* **12**, 9–17.
192. NEWSOME, A. E. (1965a). The Abundance of Red Kangaroos, *Megaleia rufa* (Desmarest), in Central Australia. *Aust. J. Zool.* **13**, 269–87.
193. NEWSOME, A. E. (1965b). The Distribution of Red Kangaroos, *Megaleia rufa* (Desmarest), about Sources of Persistent Food and Water in Central Australia. *Aust. J. Zool.* **13**, 289–99.
194. NEWSOME, A. E. (1965c). Reproduction in natural populations of the Red Kangaroo, *Megaleia rufa* (Desmarest), in Central Australia. *Aust. J. Zool.* **13**, 735–59.
195. NEWSOME, A. E. (1966). The influence of food on breeding in the Red Kangaroo in Central Australia. *C.S.I.R.O. Wildl. Res.* **11**, 187–96.
196. NEWSOME, A. E., STEPHENS, D. R. and SHIPWAY, A. K. (1967). Effect of a long drought on the abundance of red kangaroos in Central Australia. *C.S.I.R.O., Wildl. Res.* **12**, 1–8.
197. PATTERSON, B. and PASCUAL, R. (1968). Evolution of Mammals on Southern Continents. 5. The fossil mammal fauna of South America. *Q. Rev. Biol.* **43**, 409–51.
198. PEARSON, J. (1945). The female urogenital system of the Marsupialia with special reference to the vaginal complex. *Pap. Proc. R. Soc. Tasm.* **1944**, 71–98.
199. PETAJAN, J. H., MORRISON, P. R. and AKERT, K. (1962). Localization of central nervous control of temperature regulation in the opossum. *J. exp. Zool.* **150**, 225–31.
200. PETERSON, R. L. and DOWNING, S. C. (1956). Distributional records of the opossum in Ontario. *J. Mammal.* **37**, 431–35.
201. PETRIDES, G. A. (1949). Sex and age determination in the opossum. *J. Mammal.* **30**, 364–78.
202. PFEIFFER, E. W. (1968). Comparative anatomical observations of the mammalian renal pelvis and medulla. *J. Anat.* **102**, 321–31.
203. PILTON, P. E. and SHARMAN, G. B. (1962). Reproduction in the Marsupial *Trichosurus vulpecula. J. Endocr.* **25**, 119–36.
204. PLAKKE, R. K. and PFEIFFER, E. W. (1964). Blood vessels of the mammalian renal medulla. *Science, N.Y.* **196**, 1683–85.
205. PLAKKE, R. K. and PFEIFFER, E. W. (1965). Influence of plasma urea on urine concentration in the opossum (*Didelphis marsupialis virginiana*). *Nature, Lond.* **207**, 866–67.
206. PLAKKE, R. K. and PFEIFFER, E. W. (1970). Urea, electrolyte and total solute excretion following water deprivation in the opossum (*Didelphis marsupialis virginiana*). *Comp. Biochem. Physiol.* **34**, 325–32.

207. PRACY, L. T. (1962), Introduction and Liberation of the Opossum (*Trichosurus vulpecula*) into New Zealand. *Mon. Newsl. N.Z. Forest Serv.* **45**, 1–28.

208. RAVEN, H. C. and GREGORY, W. K. (1946). Adaptive branching of the kangaroo family in relation to habitat. *Am. Mus. Novit.* No. 1309, 1–33.

209. REID, I. A. and MCDONALD, I. R. (1968a). Renal function in the marsupial *Trichosurus vulpecula*. *Comp. Biochem. Physiol.* **25**, 1071–79.

210. REID, I. A. and MCDONALD, I. R. (1968b). Bilateral adrenalectomy and steriod replacement in the marsupial *Trichosurus vulpecula*. *Comp. Biochem. Physiol.* **26**, 613–25.

211. RENFREE, M. B. (1970). Protein, amino acids and glucose in the yolk-sac fluids and maternal blood sera of the tammar wallaby, *Macropus eugenii* (Desmarest). *J. Reprod. Fert.* **22**, 483–92, and **24**, 132–3.

212. REYNOLDS, H. C. (1945). Some aspects of the life history and ecology of the opossum in central Missouri. *J. Mammal.* **26**, 361–79.

213. REYNOLDS, H. C. (1952). Studies on reproduction in the opossum (*Didelphis virginiana virginiana*). *Univ. Calif. Publs. Zool.* **52**, 223–83.

214. RIDE, W. D. L. (1959). Mastication and Taxonomy in the Macropodine Skull. *System. Assoc. Publ.* No. 3, 33–59.

215. RIDE, W. D. L. (1964). A Review of Australian Fossil Marsupials. *J. Proc. R. Soc. West. Aust.* **47**, 97–131.

216. RIDE, W. D. L. (1968). On the past, present, and future of Australian Mammals. *Aust. J. Sci.* **31**, 1–11.

217. RIDE, W. D. L. and TYNDALE-BISCOE, C. H. (1962). The results of an expedition to Bernier and Dorre Islands. Mammals. *W. Australian Fish. Bull.* **2**, 54–97.

218. ROBERTS, W. W., BERGQUIST, E. H. and ROBINSON, T. C. L. (1969) Thermoregulatory grooming and sleep-like relaxation induced by local warming of preoptic area and anterior hypothalamus in opossum. *J. comp. physiol. Psychol.* **67**, 182–88.

219. ROBERTS, W. W., STEINBERG, M. L. and MEANS, L. W. (1967). Hypothalamic mechanisms for sexual, aggressive, and other motivational behaviors in the opossum, *Didelphis virginiana*. *J. comp. physiol. Psychol.* **64**, 1–15.

220. ROBINSON, K. W. (1954). Heat tolerances of Australian monotremes and marsupials. *Aust. J. biol. Sci.* **7**, 348–60.

221. ROBINSON, K. W. and MORRISON, P. R. (1957). The reaction to hot atmospheres of various species of Australian marsupial and placental animals. *J. cell. comp. Physiol.* **49**, 455–78.

222. ROMER, A. S. (1966). *Vertebrate Palaeontology.* 3rd edn. Univ. Press, Chicago.

223. ROWLANDS, D., LA VIA, M. F. and BLOCK, M. H. (1964). The Blood forming Tissues and Blood of the Newborn Opossum (*Didelphys virginiana*). 2. Ontogenesis of antibody formation to flagella of *Salmonella typhi*. *J. Immun.* **93**, 157–64.

224. ROWLANDS, D. T. and DUDLEY, M. A. (1968). The isolation of immunoglobulins of the adult opossum (*Didelphys virginiana*) *J. Immun.* **100**, 736–43.

225. SADLEIR, R. M. F. S. (1965). Reproduction in two species of Kangaroo *Macropus robustus* and *Megaleia rufa* in the Arid Pilbara region of Western Australia. *Proc. zool. Soc. Lond.* **145**, 239–61.

226. SANDERSON, G. C. (1961). Estimating opossum populations by marking young. *J. Wildl. Mgmt.* **25**, 20–27.

227. SCHAFER, E. A. and WILLIAMS, D. J. (1876). On the structure of the mucous membrane of the stomach of the Kangaroos. *Proc. zool. Soc. Lond.* **1876**, 167–77.

228. SCHMIDT-NIELSEN, K. and NEWSOME, A. E. (1962). Water balance in the mulgara (*Dasycercus cristicauda*), a carnivorous desert marsupial. *Aust. J. biol. Sci.* **15**, 683–89.

229. SCHULTZE-WESTRUM, T. G. (1969). Social communication by chemical signals in flying phalangers (*Petaurus breviceps papuanus*). In *Olfaction and Taste.* ed. Pfaffman, C. Rockefeller Univ. Press, New York.

230. SEMON, R. (1894). Die embryonalhullen der Monotremen und Marsupialie. *Denkschr. med. naturwiss. Ges. Jena.* **5**, 19–58.

231. SETCHELL, B. P. and WAITES, G. M. H. (1969). Pulse attenuation and countercurrent heat exchange in the internal spermatic artery of some Australian marsupials. *J. Reprod. Fert.* **20**, 165–69.

232. SETCHELL, P. J. (1973). The development of thermoregulation and thyroid function in the marsupial, *Macropus eugenii* (Desmarest). *Gen. comp. Endocr.* (in press).

233. SHARMAN, G. B. (1961a). The mitotic chromosomes of marsupials and their bearing on taxonomy and phylogeny. *Aust. J. Zool.* **9**, 38–60.

234. SHARMAN, G. B. (1961b). The embryonic membranes and placentation in five genera of diprotodont marsupials. *Proc. zool. Soc. Lond.* **137**, 197–220.

235. SHARMAN, G. B. (1962). The initiation and maintenance of lactation in the marsupial, *Trichosurus vulpecula. J. Endocr.* **25**, 375–85.

236. SHARMAN, G. B. (1963). Delayed Implantation in Marsupials. In *Delayed Implantation.* ed. Enders, A. C. Univ. Press, Chicago.

237. SHARMAN, G. B. (1964). The effects of the suckling stimulus and oxytocin injection on the corpus luteum of delayed implantation in the Red Kangaroo. *Excerpta med.* **83**, 669–74.

238. SHARMAN. G. B. (1965). Marsupials and the Evolution of Viviparity. *Viewpoints in Biology* **4**, 1–28.

239. SHARMAN, G. B. (1970). Reproductive physiology of marsupials. *Science, N.Y.* **167**, 1221–28.

240. SHARMAN, G. B. and BERGER, P. J. (1969). Embryonic diapause in Marsupials. *Adv. Reprod. Physiol.* **4**, 211–40.

241. SHARMAN, G. B. and CALABY, J. H. (1964). Reproductive behaviour in the Red Kangaroo, *Megaleia rufa*, in captivity. *C.S.I.R.O. Wildl. Res.* **9**, 58–85.

242. SHARMAN, G. B., CALABY, J. H. and POOLE, W. E. (1966). Patterns of Reproduction in Female Diprotodont Marsupials. *Symp. zool. Soc. Lond.* **15**, 205–32.

243. SHARMAN, G. B. and CLARK, M. J. (1967). Inhibition of Ovulation by the corpus luteum in the Red Kangaroo, *Megaleia rufa. J. Reprod. Fert.* **14**, 129–37.

244. SHARMAN, G. B., FRITH, H. J. and CALABY, J. H. (1964). Growth of the pouch young, tooth eruption and age determination in the Red Kangaroo, *Megaleia rufa. C.S.I.R.O. Wildl. Res.* **9**, 20–49.

245. SHARMAN, G. B. and PILTON, P. (1964). The life history and reproduction of the red kangaroo (*Megaleia rufa*). *Proc. zool. Soc. Lond.* **142**, 29–48.

246. SHARMAN, G. B., ROBINSON, E. S., WALTON, S. M. and BERGER, P. J. (1970). Sex chromosomes and reproductive anatomy of some intersexual marsupials. *J. Reprod. Fert.* **21**, 57–68.

247. SHIELD, J. (1961). The development of certain external characters in the young of the macropod marsupial *Setonix brachyurus*. *Anat. Rec.* **140**, 289–94.

248. SHIELD, J. (1964). A breeding season difference in two populations of the Australian macropod marsupial (*Setonix brachyurus*) *J. Mammal.* **45**, 616–25.

249. SHIELD, J. W. (1966). Oxygen consumption during pouch development of the macropod marsupial *Setonix brachyurus*. *J. Physiol. Lond.* **187**, 257–70.

250. SHIELD, J. W. and WOOLLEY, P. (1961). Age estimation by measurement of pouch young of the quokka (*Setonix brachyurus*). *Aust. J. Zool.* **9**, 14–23.

251. SHIELD, J. W. and WOOLLEY, P. (1963). Population Aspects of Delayed Birth in the Quokka (*Setonix brachyurus*). *Proc. zool. Soc. Lond.* **141**, 783–90.

252. THORBURN, G. D., COX, R. I. and SHOREY, C. D. (1971). Ovarian Steroid Secretion Rates in the Marsupial *Trichosurus vulpecula*. *J. Reprod. Fert.* **24**, 139.

253. SHORTRIDGE, G. C. (1909). An account of the geographical distribution of the marsupials and monotremes of South-West Australia, having special reference to the specimens collected during the Balston Expedition of 1904–1907. *Proc. zool. Soc. Lond.* **1909**, 803–48.

254. SIMPSON, G. G. (1944), *Tempo and Mode in Evolution*. Columbia Univ. Press, New York.

255. SIMPSON, G. G. (1961). Historical zoogeography of Australian mammals. *Evolution N.Y.* **15**, 431–46.

256. SIMPSON, G. G. (1970). The Argyrolagidae, extinct South American marsupials. *Bull. Mus. comp. Zool. Harv.* **139**, 1–86.

257. SLAUGHTER, B. H. (1968). Earliest known marsupials. *Science N.Y.* **162**, 254–55.

258. SLOAN, R. E., JENNESS, R., KENYON, A. L. and REGEHR, E. A. (1961). Comparative biochemical studies of milks. 1 Electrophoretic analysis of milk proteins. *Comp. Biochem. Physiol.* **4**, 47–62.

259. SMITH, A. G. and HALLAM, A. (1970). The fit of the southern Continents. *Nature, Lond.* **225**, 139–44.

260. SMITH, M. J. and SHARMAN, G. B. (1969). Development of dormant blastocysts induced by oestrogen in the ovariectomised marsupial, *Macropus eugenii*. *Aust. J. biol. Sci.* **22**, 171–80.

261. SMITH, R. F. C. (1969). Studies on the marsupial glider, *Schoinobates volans* (Kerr). 1. Reproduction. *Aust. J. Zool.* **17**, 625–36.

262. STIRTON, R. A., TEDFORD, R. H. and WOODBURNE, M. O. (1968). Australian Tertiary deposits containing terrestrial Mammals. *Univ. Calif. Publs. Geol.* **77**, 1–30.

263. STODART, E. (1966). Observations on the behaviour of the marsupial *Bettongia lesueuri* (Quoy and Gaimard) in an enclosure. *C.S.I.R.O. Wildl. Res.* **11**, 91–99.

264. STORR, G. M. (1964). Studies on Marsupial Nutrition. 4. Diet of The Quokka, *Setonix brachyurus* (Quoy and Gaimard), on Rottnest Island, Western Australia. *Aust. J. biol. Sci.* **17**, 469–81

265. STORR, G. M. (1968). Diet of kangaroos (*Megaleia rufa* and *Macropus robustus*) and merino sheep near Port Hedland, Western Australia. *J. Proc. R. Soc. West. Aust.* **51**, 25–32.

266. SUTHERLAND, A. (1897). The temperatures of reptiles, monotremes and marsupials. *Proc. R. Soc. Vict.* **9**, 57–67.

267. TAMAR, H. (1961). Taste reception in the opossum and the bat. *Physiol. Zool.* **34**, 86–91.

268. TAYLOR, D. L. and BURRELL, R. (1968). The immunologic responses of the North American opossum (*Didelphys virginiana*). *J. Immun.* **101**, 1207–16.

269. TEDFORD, R. H. (1955). Report on the extinct mammalian remains at Lake Menindee, New South Wales. *Rec. S. Aust. Mus.* **11**, 299–305.

270. THOMAS, O. (1888). *Catalogue of the Marsupialia and Monotremata in the collection of the British Museum (Natural History).* British Museum, London.

271. THOMSON, J. A. and OWEN, W. H. (1964). A Field Study of the Australian Ringtail Possum *Pseudocheirus peregrinus* (Marsupialia: Phalangeridae). *Ecol. Monogr.* **34**, 27–52.

272. TOMICH, P. Q. (1969). *Mammals in Hawaii.* Bishop Museum Press, Honolulu.

273. TRIBE, D. E. and PEEL, L. (1963). Body composition of the kangaroo (*Macropus sp.*). *Aust. J. Zool.* **11**, 273–89.

274. TYNDALE-BISCOE, C. H. (1963a). The role of the corpus luteum in delayed implantation in marsupials. In *Delayed Implantation.* ed. Enders, A. C. Univ. Press, Chicago.

275. TYNDALE-BISCOE, C. H. (1963b). Effects of ovariectomy in the marsupial, *Setonix brachyurus.* *J. Reprod. Fert.* **6**, 25–40.

276. TYNDALE-BISCOE, C. H. (1963c). Blastocyst transfer in the marsupial *Setonix brachyurus.* *J. Reprod. Fert.* **6**, 41–48.

277. TYNDALE-BISCOE, C. H. (1966). The Marsupial Birth Canal. *Symp. zool. Soc. Lond.* **15**, 233–50.

278. TYNDALE-BISCOE, C. H. (1968). Reproduction and post-natal development in the marsupial *Bettongia lesueur* (Quoy and Gaimard). *Aust. J. Zool.* **16**, 577–602.

279. TYNDALE-BISCOE, C. H. (1969). Relaxin activity during the oestrous cycle of the marsupial, *Trichosurus vulpecula* (Kerr). *J. Reprod. Fert.* **19**, 191–93.

280. TYNDALE-BISCOE, C. H. (1970). Resumption of development by quiescent blastocysts transferred to primed, ovariectomised recipients in the marsupial, *Macropus eugenii.* *J. Reprod. Fert.* **23**, 25–32.

281. TYNDALE-BISCOE, C. H. and SMITH, R. F. C. (1969a). Studies on the marsupial glider *Schoinobates volans* (Kerr). 2. Population structure and regulatory mechanisms. *J. Anim. Ecol.* **38**, 637–50.

282. TYNDALE-BISCOE, C. H. and SMITH, R. F. C. (1969b). Studies on the marsupial glider *Schoinobates volans* (Kerr). 3 Response to habitat destruction. *J. Anim. Ecol.* **38**, 651–59.

283. TYNDALE-BISCOE, C. H. and WILLIAMS, R. M. (1955). A study of natural mortality in a wild population of the rabbit, *Oryctolagus cuniculus* (L.). *N.Z. Jl. Sci. Technol.* **36**, 561–80.

284. UREN, J. and MOORE, R. (1966). Permanent cell lines of the marsupial mouse *Antechinus swainsonii.* *Expl. Cell. Res.* **44**, 273–82.

285. WALEN, K. H. and BROWN, S. W. (1962). Chromosomes in a marsupial (*Potorous tridactylus*) tissue culture. *Nature, Lond.* **194**, 406.

286. WATSON, D. M. S. (1917). The evolution of the tetrapod shoulder girdle and forelimb. *J. Anat.* **52**, 1–63.
287. WEISS, M. and MCDONALD, I. R. (1966). Corticoid secretion in the Australian phalanger (*Trichosurus vulpecula*). *Gen. comp. Endocr.* **7**, 345–51.
288. WEISS, M. and MCDONALD, I. R. (1967). Corticosteroid secretion in Kangaroos (*Macropus Canguru major*) and *M. (Megaleia) rufus*. *J. Endocr.* **39**, 251–61.
289. WHITE, A., HANDLER, P. and SMITH, E. L. (1964). *Principles of Biochemistry*. 3rd. ed. McGraw-Hill, New York.
290. WISEMAN, G. L. and HENDRICKSON, G. O. (1950). Notes on the life history and ecology of the opossum in south east Iowa. *J. Mammal.* **31**, 331–37.
291. WODZICKI, K. A. (1950). Introduced mammals of New Zealand. *Bull. N.Z. Dep. scient. ind. Res.* No. 98.
292. WOOD, D. H. (1970). An ecological study of *Antechinus stuartii* (Marsupialia) in a South-east Queensland rain forest. *Aust. J. Zool.* **18**, 185–207.
292a. WOOLLARD, P. (1971). Differential mortality of *Antechinus stuartii* (Macleay): nitrogen balance and somatic changes. *Aust. J. Zool.* **19**, 347–53.
293. WOOLLEY, P. (1966). Reproduction in *Antechinus* spp. and other Dasyurid Marsupials. *Symp. zool. Soc. Lond.* **15**, 281–94.
294. YADAV, M. and PAPADIMITRIOU, J. M. (1969). The ultrastructure of the neonatal thymus of a marsupial, *Setonix brachyurus*. *Aust. J. exp. Biol. med. Sci.* **47**, 653–68.
295. YAPP, W. B. (1970). *The Life and Organisation of Birds*. Arnold, London.

Index

Italicized page numbers refer to Figures